ROUTLEDGE LIBRARY EDITIONS:
TRANSPORT ECONOMICS

Volume 3

AIRLINE OPERATIONS RESEARCH

AIRLINE OPERATIONS RESEARCH

DUŠAN TEODOROVIĆ

Routledge
Taylor & Francis Group
LONDON AND NEW YORK

First published in 1988 by Gordon and Breach Science Publishers

This edition first published in 2017
by Routledge
4 Park Square, Milton Park, Abingdon, Oxon OX14 4RN
605 Third Avenue, New York, NY 10017

Routledge is an imprint of the Taylor & Francis Group, an informa business

British Library Cataloguing in Publication Data
A catalogue record for this book is available from the British Library

ISBN: 978-0-415-78484-9 (Set)
ISBN: 978-1-315-20175-7 (Set) (ebk)
ISBN: 978-0-415-79224-0 (Volume 3) (hbk)
ISBN: 978-1-315-21184-8 (Volume 3) (ebk)

Publisher's Note
The publisher has gone to great lengths to ensure the quality of this reprint but points out that some imperfections in the original copies may be apparent.

Disclaimer
The publisher has made every effort to trace copyright holders and would welcome correspondence from those they have been unable to trace.

Airline Operations Research

By

Dušan Teodorović

GORDON AND BREACH SCIENCE PUBLISHERS
New York • London • Paris • Montreux • Tokyo • Melbourne

Gordon and Breach Science Publishers

Post Office Box 786
Cooper Station
New York, New York 10276
United States of America

58, rue Lhomond
75005 Paris
France

3-14-9, Okubo
Shinjuku-ku, Tokyo
Japan

Post Office Box 197
London WC2E 9PX
England

Post Office Box 161
1820 Montreux 2
Switzerland

Private Bag 8
Camberwell, Victoria 3124
Australia

Library of Congress Cataloging-in-Publishing Data

Teodorovic. D. (Dušan). 1951-
Airline operations research / by Dušan Teodorovic.
p. cm. — (Transportation studies ; v. 10)
Translated from Serbo-Croatian (Cyrillic).
Bibliography:
Includes index.
ISBN 2-88124-672-9
1. Aeronautics, Commercial. 2. Operations research. I. Title.
TL552.T46 1988
367.7'068-dc19 8-21275

CIP

Contents

Introduction to the Series

This broad-ranging series of books will cover many of the varied aspects of transportation. The subject area will be generally divided into two parts: the first deals with planning and technological aspects of transportation, the second with specialized transportation.

Within the context of this series, technology and planning will include the wide spectrum of various aspects of the design and planning of vehicles and infrastructure for the transport of freight and passengers, as well as operational and management considerations. The general aim of the planning and technology series is to inform readers of the state of the art and to summarize the status of transportation.

The second part of the series will seek to generate monographs dealing with the mobility of those groups in society increasingly characterized as the transportation disadvantaged, particularly, but not exclusively, the elderly, the disabled and families with low incomes. It is anticipated that the content of these books will be derived largely from research, policy analysis and documental field experience. The subject matter will include advances in the relevant technology, service and methods demonstration, improved planning and methodology, major or proposed changes in public policy and innovative proposals for the development or change of systems.

Occasionally, more specific monographs will be published, presenting the results of individual studies in areas of special interest to planners and technologists.

As with any monograph series, the emphasis is on current information, and the material will be of interest to the transport practitioner, the postgraduate student and the academic working in the field.

NORMAN ASHFORD

WILLIAM G. BELL

Preface

During the past several decades, air transportation has undergone very intensive growth. With this development appeared an entire range of practical problems to be solved. Successfully solving these problems required operations research methods and modern computer technology, whose development and application were primarily the work of specialized air carrier departments. The annual AGIFORS symposiums made a considerable contribution to air carriers' exchanges of experience and to further development of operations research methods in the field of air transportation.

Air carrier problems are also of interest to some universities. When writing Airline Operations Research my primary endeavor was to facilitate students studying this field. The book is also intended for experts who encounter complex air transportation problems on a daily basis : it contains the results of works published in international scientific magazines during the past three decades and research reports from several airline companies. The main objective of the book is to make a modest contribution to the further development and application of operations research methods in air transportation through a systematization of existing knowledge.

I would like to take this opportunity to thank Alice Copple-Tosic for her translation, and Snezana and Mirko Licitar for their technical preparation of the manuscript for printing.

D. Teodorović

1. TRANSPORTATION DEMAND MODELS

1.1. Air Transportation Demand

Every day an extremely large number of trips are made and considerable amounts of merchandise are transported by air. The need for transportation arises from man's various activities. By monitoring the number of passengers flown on different routes during a day, week and month, certain patterns are noted that characterize the demand for air transportation services. In addition, air transportation developments in the past revealed certain dependencies between the demand for air transportation services and the socio-economic characteristics of the region linked by air transportation. Demand is greater in more developed regions on the one hand, and air traffic accelerates the development of the regions it links, on the other. The noted dependencies between the demand for air transportation services and the socio-economic characteristics of the region are used in the air transportation planning process. This process entails the planning of airports, needed transportation facilities, route networks and planning the network of airways.

The input data for planning route networks, transportation facilities and for building new and reconstructing existing airports includes data that characterize the demand for air transportation services. This chapter will discuss in detail the problem of transportation demand in air traffic.

Passenger demand intensity comprises the number of passengers in a unit of time that travel from one city to another. There is also a potential passenger demand intensity that includes the number of passengers in a unit of time that would like to travel from one city to another. Real and potential passenger demand intensities most often differ due to numerous factors. It is important at the very beginning of our discussion of transportation demand models to make a distinction between effectuated transportation and the demand for transportation on a certain route. Since passenger traffic is crucially important to the dimensions given to certain facilities at an airport, and since the establishment of air traffic between two cities is conditioned by the demand for transportation between those two cities, a precise evaluation must be given of the passenger demand between pairs of cities and the passenger traffic at different airports.

1.2. The Supply of Air Transportation Services

The supply of air transportation services normally comprises the set of following attributes : travel time, waiting time at the airport, transportation tariffs, ground costs, flight frequency, airline schedule, airplane type and cabin service.

Travel time between two cities most often comprises the total time needed for a passenger to go from the center of the city of departure to the center of another city. This means that travel time to and from the airport, waiting time and transfer time must also be added to flight time. Travel time is a factor that considerably affects the passenger's decision to use a certain mode of transportation. Travel time from city center to city center should be used when discussing how passengers choose a mode of transportation and to estimate the share of air transportation out of the total number of trips on a certain route.

Travel costs normally comprise the passenger's total costs. This includes, in addition to the cost of the ticket, all other passenger outlays during the trip.

Flight frequency and airline schedule are extremely important attributes to the supply of air transportation services. Flight frequency is the number of flights that take place on a route during a fixed period of time. Flight frequency most often comprises the daily frequency of flights or the number of flights per day on a specific route. The quality of air transportation services certainly depends on flight frequency and the size of the aircraft providing transportation. The airline schedule, or the distribution of airplane departures, also has an essential effect on the quality of supply. The airline schedule is adjusted more or less to passenger demand and this adjustment is usually measured by the average difference (per passenger) between the time in which the passenger wants to depart and the time in which this is really possible due to the existing airline schedule.

Comfort during the trip, i.e. cabin service, is evaluated by polling the passengers. Airline carriers take surveys of the passengers from time to time thereby establishing the quality of cabin service.

The supply of transportation service has an essential effect on air transportation demand. For example, by increasing flight frequency, better adjusting the airline schedule to passenger requests, reducing tariffs or some other promotional measure, air transportation demand can be increased to a certain extent and thereby the number of passengers flown on certain routes. In the same vein, a decline in demand is often caused by an inadequate supply of transportation facilities and poor transportation services offered to passengers. Due to this dependence between transport demand and supply, many models that evaluate transportation demand have built in certain attributes of the supply of transportation services in both air transportation and in other competitive modes of transportation on a specific route.

1.3. <u>A Classification of Demand Estimation Models</u>

The previous presentation underscored that demand for transportation between two cities or two regions depends on both the socio-economic characteristics of the regions in question and on the characteristics of the transportation system that links them. Models to evaluate air transportation demand most often evaluate the number of potential passengers, the number of passenger kilometers that can be achieved, the expected number of operations (take offs and landings) or a percentage share of the number of air passengers out of the total number of passengers. Some of these quantities are usually presented as dependent variables in the model. Independent variables are presented in the model as certain socio-economic characteristics and/or characteristics of the transportation system. The process of forecasting demand for air transportation services comprises above all making a precise definition of the problem that gives rise to the need for planning, and then forecasting certain socio-economic characteristics. For example, let there be a city in which we wish to build an airport. We are presented with the problem of justifying the construction of this airport. In order to answer this question, we start with a planning process whose output should help us in reaching an appropriate decision. The next phase is usually the creation of suitable statistical data. This means surveying automobile drivers, passengers at bus and train stations, households and hotel guests, to get data on origin and destination trends, on business and private trips, on departures for vacation, etc. If, for example, we want to open a new route in long-haul traffic, data should be collected on the number of inhabitants in the regions in question, national income, the volume of trade, number of tourists and all other data that might be of use. When collecting data, this should relate to longer periods of time in order to get an idea of the dynamics of certain phenomena that affect the potential number of passengers. The next step is usually to estimate the number of inhabitants, employment, national income, etc. This is an essential planning step since transportation demand is a function of certain socio-economic characteristics.

Forecasting transportation demand is certainly the most important step in transportation planning. In this phase, depending on the context of the problem being discussed, it is possible to forecast the total number of trips, the number of trips between specific zones, the percentage share of individual modes of transportation in the total number of trips between specific zones, and the number of trips by certain modes of transportation along the routes joining these zones. The process of forecasting transportation demand most often comprises the following steps :

a) trip generation,

b) trip distribution,

c) modal split,

d) trip assignment.

These four planning phases are the normal planning methodology that is applied when elaborating a large number of transportation plans for different regions. However, depending on the context of the problem under consideration, the demand forecast problem does not have to include all four of the above planning phases.

Trip generation comprises an estimate of the total number of future trips from the region of a certain zone (Fig. 1.1a). When distributing trips, the number of trips between individual zones is established (Fig. 1.1b).

Modal split comprises an estimation of the number of trips by individual types of transportation mode (automobile, train, bus, airplane) (Fig. 1.1c).

Upon establishing the modal split, trips are usually assigned to routes (Fig. 1.1d).

Two points can be linked by numerous possible routes.

Trip assignment comprises an estimate of the expected number of trips along individual routes.

When speaking of models to estimate demand in air transportation, it should be underscored that not all of the above-mentioned planning processes must always be used, i.e. the phases of trip generation, trip distribution, modal split and trip assignment. The types of models that will be applied in order to estimate air transportation demand primarily depend on the nature of the problem under consideration. In certain cases transportation demand should be estimated independently of the other types of transportation. In other cases transportation demand must be estimated by taking into consideration the characteristics of competitive transportation modes. Depending on whether or not the model includes competitive modes of transportation, demand estimation models in air transportation can be divided into :

a) multimode models,

b) estimated demand models that are independent of the characteristics of alternative modes of transportation.

In long haul traffic, different modes of transportation are most often studied independently of each other since the longer the trip the smaller the number of alternative modes of transportation. The airplane is the predominant mode of transportation on many long-distance traffic routes. Travel time is much more important to long-distance

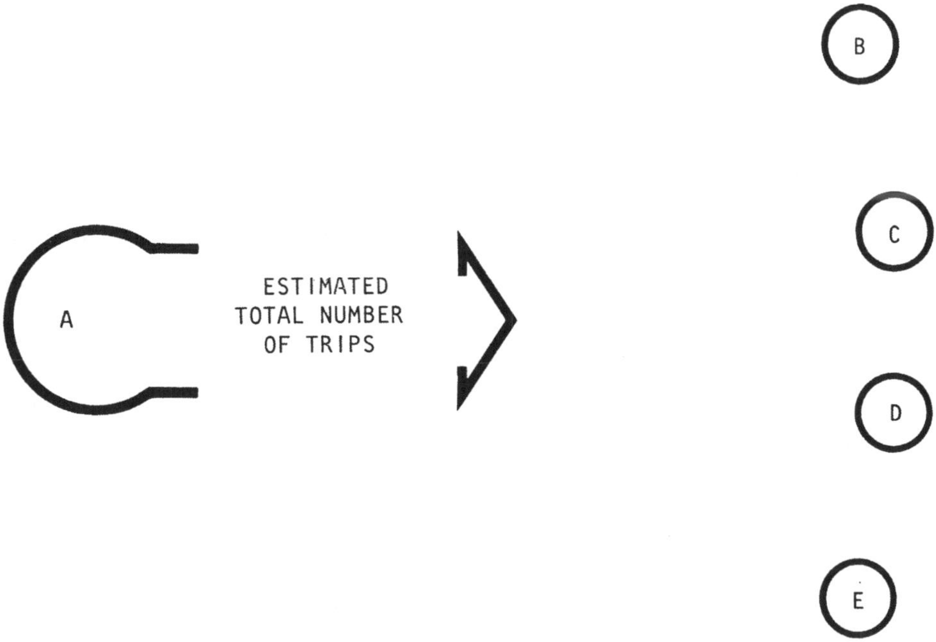

FIGURE 1.1a. Estimated total number of trips from a zone area

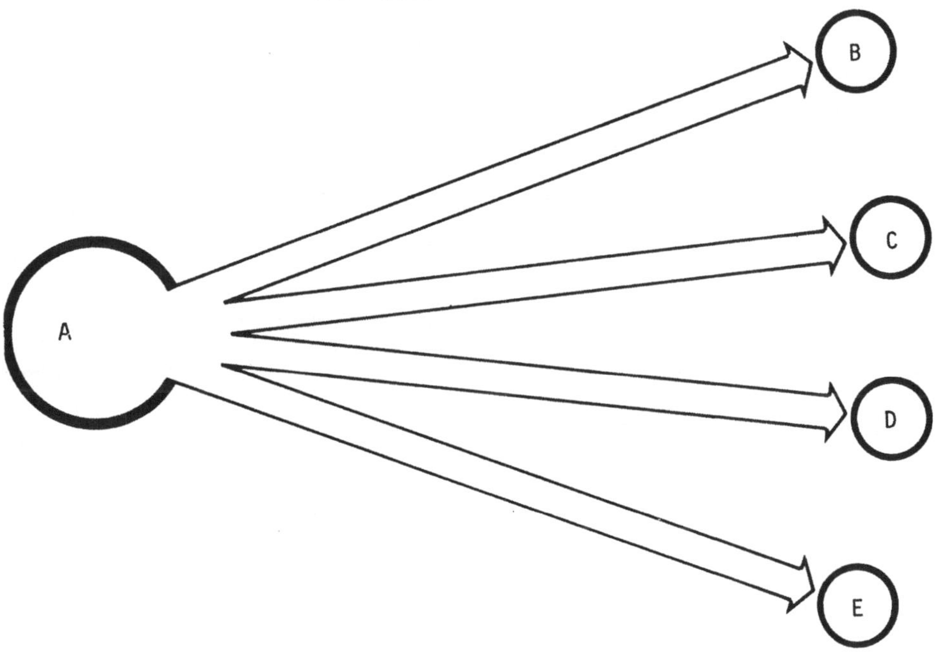

FIGURE 1.1b Trip distribution

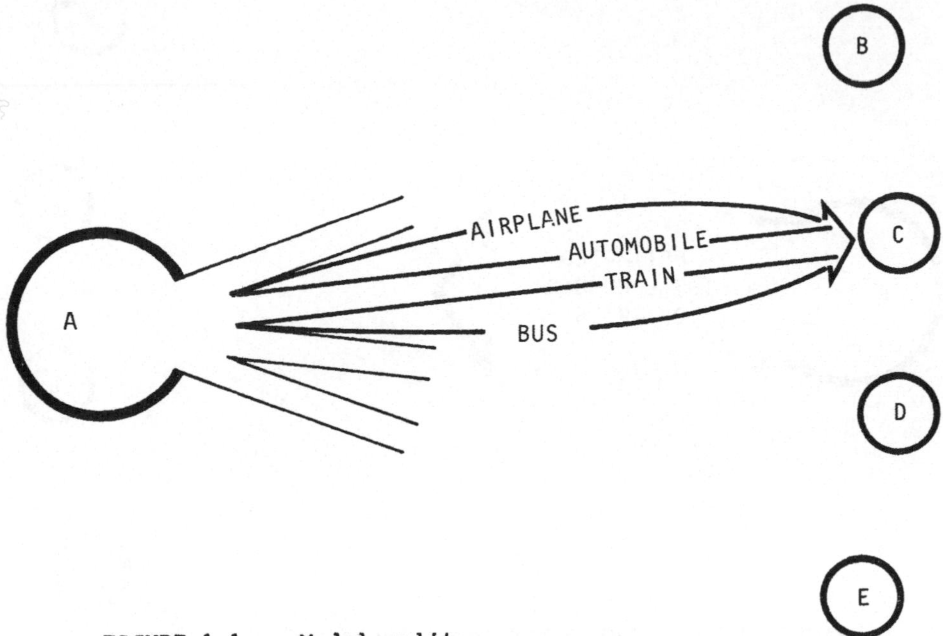

FIGURE 1.1c   Modal split

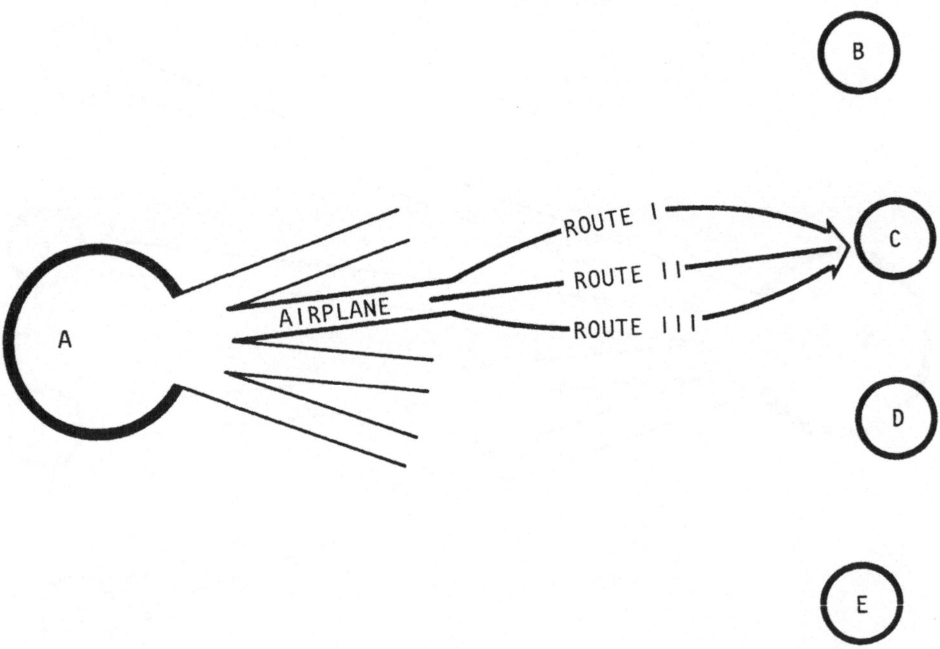

FIGURE 1.1d   Trip assignment

passengers than the price of the ticket. Therefore, demand for air transportation on long-haul routes should be estimated independently of other modes of transportation. This demand is estimated on the basis of yearly and monthly data concerning the number of passengers transported, under the assumption that no essential changes will take place in traffic conditions.

Multimode models are primarily used to estimate demand for air transportation on short-haul routes. The presence of numerous competitive modes of transportation and a greater number of possible routes complicates the estimation of potential passenger flows. Since travel time is relatively short, passengers in short-haul traffic are very sensitive to the proffered supply of daily departures of different modes of transportation and to the level to which bus, train and airline schedules are adjusted to passenger requests in terms of travel time. In addition, passengers making personal trips are particularly sensitive to the price of transportation. Therefore, air transportation demand on shorter routes is usually estimated simultaneously with the estimation of demand on other modes of transportation.

As noted by Kanafani[74] models of air transportation demand can be divided into :

a) macroscopic models, and

b) microscopic models.

Macroscopic models are used to estimate the development level of air transportation in a certain country or region. These models are used to estimate the number of passengers, the number of airplane operations or the number of passenger kilometers. The number of passengers normally comprises the total number of passengers on commercial flights in a region during a certain period of time, while the number of operations comprises the number of take offs on commercial flights. The number of passenger kilometers is arrived at by multiplying the number of trips by the length of each trip. The number of passenger kilometers is an indicator of effectuated transportation.

Microscopic models estimate air transportation demand between two cities, the passenger traffic at an airport, the number of passengers along a specific route when there are several different routes and the number of passengers in each class when there are different tariffs on a route.

1.4. Macroscopic Models of Transportation Demand When Demand is a Function of Time

The previous presentations pointed out the interrelationship between transportation demand and the socio-economic characteristics of a region. However, in certain cases an extrapolation is simply made of trends noted by the number of passengers transported or the number of passenger kilometers effectuated, and a forecast on this basis is made of the number of passengers or passenger kilometers in the next period. This is known as the independent estimation method since factors that affect the number of passengers are not taken into consideration when estimating the number of air passengers. A time series analysis can be made when using this method, which means studying the basic development tendencies of the number of air passengers, studying the cyclical fluctuation of the number of passengers over a longer period of time, studying seasonal fluctuations (which is particularly important for routes dominated by the tourist trade) and studying the random fluctuation of the number of air passengers. When studying the number of air passengers as a function of time, the correlation theory can be used in addition to time series analyses. This theory enables certain functional links to be discovered between the number of air passengers and time.

We denote time by t , and the number of air passengers that changes over time by y . The following functional links are most often encountered between the number of air passengers y and time t :

$$y = kt + m \tag{1.1}$$

In this case, it is assumed that the growth in demand for air transportation is constant within a unit of time. Model calibration, i.e. estimating parameters k and m is made using the least squares method.

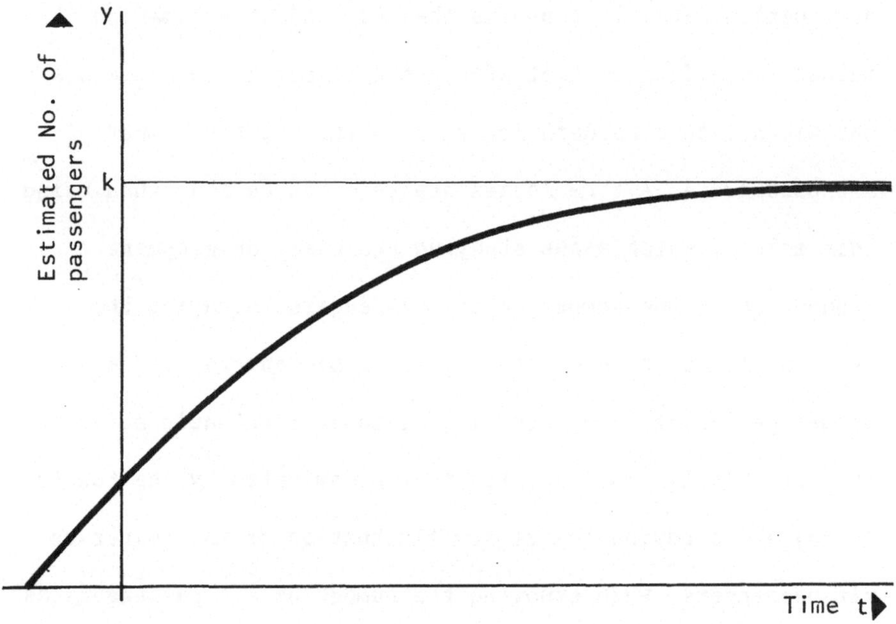

FIGURE 1.2. Modified exponential curve describing the number of air passengers as a function of time

This function is also encountered :

$$y = a \cdot b^t \tag{1.2}$$

or, in the logarithmic form :

$$\log y = \log a + t \cdot \log b \tag{1.3}$$

When transposing into the logarithmic form, parameters a and b can also be estimated using the least squares method.

The modified exponential curve has the shape :

$$y = k + a \cdot b^t \qquad (1.4)$$

A modified exponential curve where $a<0$ and $b<1$ is shown on Figure 1.2.

As can be concluded from Figure 1.2, it is assumed when applying the modified exponential curve that the future number of air passengers approaches a fixed saturation level.

The Gompertz curve takes the form :

$$y = k \cdot a^{b^t} \qquad (1.5)$$

or the logarithmic form :

$$\log y = \log k + b^t \cdot \log a \qquad (1.6)$$

The Gompertz curve is shown on Figure 1.3 for $\log a < 0$ and $b < 1$.

The Gompertz curve is also characterized by a saturation level (parameter k). The number of passengers first rises slowly. The middle section of the curve shows rapid growth and at the end the number of passengers approaches the saturation level.

The Pearl-Reed curve, also called the logistic curve, has a shape similar to the Gompertz curve. The equation for

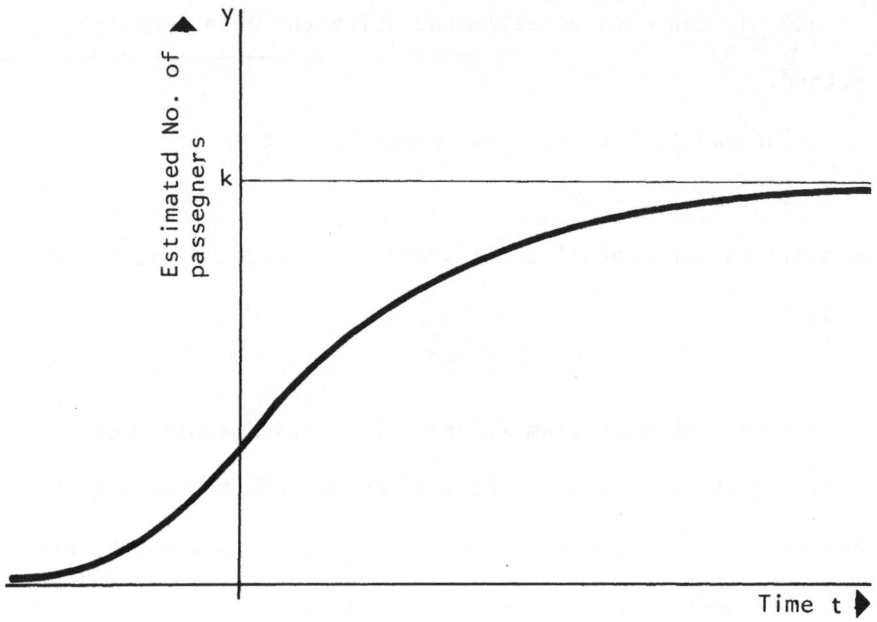

FIGURE 1.3 Gompertz curve describing the number of air passengers as a function of time.

this curve is :

$$y = \frac{k}{1 + b\, e^{-at}} \qquad (1.7)$$

The logistic curve is shown on Figure 1.4.

The least squares method cannot be applied to estimate the parameters of modified exponential curves, Pearl-Reed curves and Gompertz curves, for it produces transcendental equations that are very inconvenient to solve. The three-sum and three-point methods have proven very successful in

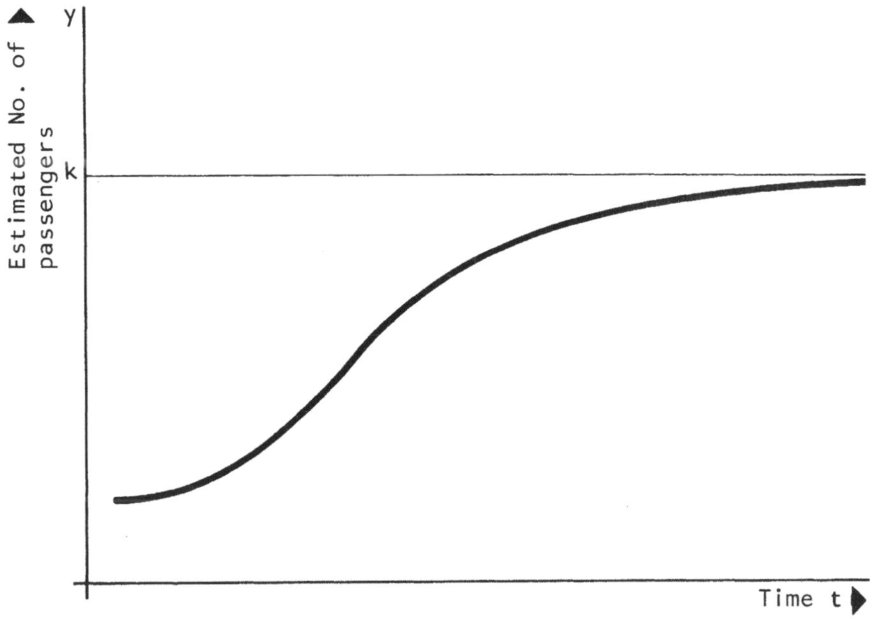

FIGURE 1.4 Logistic curve describing the number of air passengers as a function of time

estimating the parameters of these curves.

Application of the three-point method to define the parameters of a logistic curve is as follows. Three points are chosen at an equal distance from each other. Let the distance between the points be n time units and let the first point be on the curve section where passenger growth is small, the second on the middle section where the number of passengers rises rapidly, and the third on the curve section that approaches the saturation level. The ordinates of these points are denoted respectively by y_0, y_1 and y_2. For the first point, we take $t=0$. This means that :

$$y_0 = \frac{k}{1+be^{a\cdot o}} = \frac{k}{1+b} \tag{1.8}$$

$$y_1 = \frac{k}{1+be^{a\cdot n}} = \frac{k}{1+be^{a\cdot n}} \tag{1.9}$$

$$y_2 = \frac{k}{1+be^{a\cdot 2n}} = \frac{k}{1+be^{2an}} \tag{1.10}$$

We now introduce quantities d_1 and d_2 that are defined as :

$$d_1 = \frac{1}{y_o} - \frac{1}{y_1} \tag{1.11}$$

$$d_2 = \frac{1}{y_1} - \frac{1}{y_2} \tag{1.12}$$

It is shown that :

$$\frac{d_2}{d_1} = e^{na} \tag{1.13}$$

and :

$$a = \frac{1}{n}\left[ln\ d_2 - ln\ d_1\right] \tag{1.14}$$

It is also shown that :

$$\frac{d_1^2}{d_1 - d_2} = \frac{b}{k} \tag{1.15}$$

Since :

$$\frac{1}{y_o} = \frac{1}{k} + \frac{b}{k} \tag{1.16}$$

then :

$$k = \frac{1}{\dfrac{1}{y_o} - \dfrac{d_1^2}{d_1 - d_2}} \tag{1.17}$$

And since :

$$\frac{1}{y_o} = \frac{1 + b}{k} \qquad (1.18)$$

then :

$$b = \frac{k - y_o}{y_o} = \frac{\dfrac{1}{\dfrac{1}{y_o} - \dfrac{d_1^2}{d_1 - d_2}} - y_o}{y_o} \qquad (1.19)$$

Grilihes [62] describes total passenger traffic at the Belgrade Airport as a function of time using a logistic curve. Data on passenger traffic from 1962-1978 are given in Table 1.1.

TABLE 1.1 Data on passenger traffic at the Belgrade Airport from 1962-1978

Year	*Total passenger turnover at Belgrade airport*	*Year*	*Total passenger turnover at Belgrade airport*
1962	220.726	1970	838.156
1963	282.873	1971	1.036.311
1964	329.619	1972	1.155.166
1965	405.191	1973	1.434.454
1966	335.999	1974	1.688.247
1967	399.066	1975	2.020.291
1968	462.919	1976	2.047.016
1969	602.257	1977	2.280.972
		1978	2.660.670

By applying the three-point method, a logistic equation is

obtained that can be used to describe passenger traffic at the Belgrade airport from 1962-1978. This equation reads :

$$y = \frac{9.023.394}{1+39,88\cdot e^{-0,176\cdot t}} \tag{1.20}$$

Bjorkman [15a] developed a model to estimate the number of future passengers on routes that have existed for many years. The number of future passengers is defined by the equation :

$$T_i = T_s \left(\frac{y_i - y_o}{y_s - y_o} \right)^2 \tag{1.21}$$

where :

T_i - the forecasted number of passengers in year y_i,

T_s - the number of passengers in year y_s when forecasting started,

y_i - the year being forecasted for passenger number,

y_s - the year when forecasting started,

y_o - the year traffic started on the route in question.

The established functional link between the number of air passengers and time in the past is used to forecast the number of future air passengers, i.e. the estimated future transportation demand. Since previously described models estimated the number of future passengers exclusively on the basis of past passengers, these models have relatively limited application. Transportation demand, as already mentioned several times, is a function of specific socio-economic characteristics and the characteristics of the transportation system. In addition, the volume of transportation on certain routes is often conditioned by inter-state and inter-company agreements. A considerable amount of different information is therefore needed when estimating future transportation demand. Models that use exclusively the number of past air passengers to estimate the number of future air passengers can be used primarily to estimate demand between regions that have had air traffic between them for many years, to the effect that no essential changes can be expected in factors affecting transportation demand.

1.5 Macroscopic Models of Transportation Demand When Demand is a Function of Socio-Economic Characteristics

For macroscopic models of transportation demand when demand is a function of socio-economic characteristics, the number of passengers, the number of operations or the number of passenger kilometers are dependent variables. Independent variables are chosen from the set of socio-economic characteristics and characteristics of the transportation system. The socio-economic characteristics that are most often chosen include population, national income, personal consumption, volume of trade, number of tourist, etc. The cost of transportation and/or speed, i.e. travel time, are characteristics of the transportation system that can also be independent variables.

Macroscopic models make it possible to estimate the development level of air traffic in a specific country or region. In order to successfully calibrate macroscopic models, a good statistical-documentational base must be available. Macroscopic models where demand is a function of socio-economic characteristics and the characteristics of the transportation system can be written in the following general form :

$$Y_t = a \prod_{i=1}^{m} S_{it}^{b_i} \prod_{j=1}^{n} T_{jt}^{c_j} \qquad (1.22)$$

where :

m - the total number of socio-economic characteristics,

n - the total number of transportation system characteristics,

Y_t - the number of air passengers in year t ,

S_{it} - the value of the i-th socio-economic characteristic in year t ,

T_{jt} - the value of the j-th transportation system

characteristic in year t,

a, b_i, c_j - parameters that are estimated statistically.

By finding the logarithm of the left and right sides of equation (1.22) we get the following :

$$\log Y_t = \log a + \sum_{i=1}^{m} b_i \log S_{it} + \sum_{j=1}^{n} c_j \log T_{jt} \qquad (1.23)$$

Parameters a, b_i and c_j can now be estimated by applying the multiple regression technique. These parameters can also be estimated by the maximum likelihood function. When this function is used there is no need for the model to take a preliminary linear form by finding the logarithm.

TABLE 1.2 Number of tourists and charter passengers in both directions between Great Britain and Yugoslavia from 1969-1977

Year t	No.of tourists in both directions between great Britain and Yugoslavia S_t	No. of passengers on charter flights in both directions between great Britain and Yugoslavia Y_t
1969	282.000	124.816
1970	285.000	153.156
1971	325.400	177.642
1972	314.709	184.163
1973	378.816	239.163
1974	266.755	193.210
1975	338.726	210.000
1976	381.286	235.000
1977	246.063	123.000

Table 1.2 presents the number of tourists in both directions and the number of charter passengers in both directions between Great Britain and Yugoslavia from 1969-1977.

Grulovic[63] demonstrated that a functional link of the following form exists between Y_t and S_t :

$$Y_t = 1,734 \cdot S_t - 353.328 \tag{1.24}$$

where the correlation coefficient is r = 0.914.

The number of passengers on regular flights was analyzed for the same period of time. Table 1.3 presents the trade

TABLE 1.3 The trade volume and the number of passengers on regular flights in both directions between Great Britain and Yugoslavia from 1969-1977

Year t	*Trade in million dinars* S_t	*No.of passengers on regular flights in both directions* Y_t
1969	3,474	25,534
1970	4,680	42,568
1971	4,865	46,722
1972	4,533	44,236
1973	5,426	63,900
1974	5,482	71,582
1975	6,599	94,802
1976	7,502	107.090
1977	7,668	104.758

volume and the number of passengers on regular flights in both directions between Yugoslavia and Great Britain from 1969-1977.

Grulovic also confirmed that there is a functional link between the number of passengers on regular lines in both directions Y_t and the volume of trade S_t between Great Britain and Yugoslavia :

$$Y_t = 25{,}57 \cdot S_t - 57.928 \qquad (1.25)$$

where the coefficient obtained is r = 0.9866.

Behbehani and Kanafani[13], in analyzing the traffic between North America and the South Pacific from 1970-1976 established the following functional dependence between the number of passengers, income, the trade volume and transportation tariffs :

$$Y_t = e^{-12{,}6} \cdot S_{1t}^{2{,}99} \cdot S_{2t}^{0{,}2} \cdot e^{-0{,}005 \cdot T_{1t}} \qquad (1.26)$$

where :

Y_t - the number of trips in year t ,

S_{1t} - the income in year t ,

S_{2t} - the trade volume in year t ,

T_{1t} - the transportation tariff in year t .

Brown and Watkins[17] analyzed domestic air traffic in the USA from 1946-1966 and proposed the following model to estimate the yearly number of per capita passenger-miles :

$$\Delta logT = K + a\Delta logF + b\Delta logY + clogt \qquad (1.27)$$

where :

T - the effectuated number of per capita passenger miles,

F - the average cost of transportation per mile,

Y - national per capita income,

t - time,

k, a, b, c - parameters that are statistically estimated,

Δlog - the difference between logarithms in two consecutive years (e.g. $\Delta logT = logT_n - logT_{n-1}$).

After calibrating the model, Brown and Watkins obtained:

$$k = 0{,}0725; \quad a = -1{,}307; \quad b = 1{,}119; \quad c = -0{,}038$$

We have presented several macroscopic models to estimate the level of air traffic. These models are used when developing transportation studies of a specific region or country, and their primary purpose is to provide a global perception of future air transportation developments. A preliminary forecast must be made of socio-economic characteristics and transportation system characteristics that are independent variables in the model in order to forecast the number of passengers, number of operations or passenger kilometers. The quality of the forecasted air traffic depends above all on the soundness of the forecasted socio-economic characteristics, transportation system characteristics and the soundness of the assumption that the established functional dependence will be stable over time.

1.6. Trip Distribution Models

When the total number of trips that a region can generate has been established, the trips are then distributed. This distribution establishes the number of trips between individual zones. Trip distribution models are primarily applied to the problem of planning city traffic and transportation. However, these models can also be used when studying intercity transportation.

Let there be m trip origins and n trip destinations.

		1	2	3	...	n
	1	f_{11}	f_{12}	f_{13}	...	f_{1n}
	2	f_{21}	f_{22}	f_{23}	...	f_{2n}
	3	f_{31}	f_{32}	f_{33}	...	f_{3n}
TRIP ORGINS	.	.	.	.		.
	.	.	.	.		.
	.	.	.	.		.
	m	f_{m1}	f_{m2}	f_{m3}	...	f_{mn}

We denote the number of trips from origin i to destination j by f_{ij} $(i=1,2,\ldots,m;\ j=1,2,\ldots,n)$. In air traffic, $m = n$, since flights between two cities generally go in both directions. We denote by a_i the number of trips generated by node i, and by b_j the number of trips attracted to node j, so that :

$$a_i = \sum_{j=1}^{n} f_{ij} \tag{1.28}$$

$$b_j = \sum_{i=1}^{m} f_{ij} \tag{1.29}$$

The total number of trips ν made from all origins to all destinations is :

$$\nu = \sum_{i=1}^{m} a_i = \sum_{j=1}^{n} b_j \tag{1.30}$$

Trip distribution models often use u_i , ν_j and p_{ij} instead of a_i , b_j and f_{ij} , defined in the following manner :

$$u_i = \frac{a_i}{\nu} \tag{1.31}$$

$$\nu_j = \frac{b_j}{\nu} \tag{1.32}$$

$$p_{ij} = \frac{f_{ij}}{\nu} \tag{1.33}$$

It is clear that :

$$\sum_{i=1}^{m} u_i = \sum_{j=1}^{n} \nu_j = \sum_{i=1}^{m} \sum_{j=1}^{n} p_{ij} = 1 \tag{1.34}$$

We denote by c_{ij} the trip "costs" from origin i to node j . Depending on the context of the problem being considered, c_{ij} can represent the transportation tariff, travel time, a combination of financial outlays and time or some other value. The general form of the trip distribution model is :

$$f_{ij} = f(A_i, B_j, C_{ij}) \tag{1.35}$$

where :

f_{ij} - number of passengers from origin i to destination j ,

A_i - the function of the socio-economic characteristics of trip origin i ,

B_j - the function of the socio-economic characteristics of trip destination j ,

C_{ij} - the function representing the "resistance" when travelling from i to j (e.g., the "resistance" is greater the greater the transportation tariff).

We can conclude that trip distribution models are also founded on the assumption that transportation demand is a function of socio-economic characteristics and transportation system characteristics.

1.6.1 Entropy trip distribution models. Let us observe m trip origins and n trip destination in intercity traffic. We denote by p_{ij} the probability that a trip will be made from origin i to destination j . Analogous to definitions in thermodynamics and the information theory, the entropy of the transportation network can be defined as :

$$H = - \sum_{i=1}^{m} \sum_{j=1}^{n} p_{ij} \ln p_{ij} \qquad (1.36)$$

This means that an undefined trip origin and destination can be added to the knowledge that a trip took place in the network. Should we randomly "choose" the origin and destination of a trip, we can define the entropy of the transportation network as the measure of indetermination of the experiment comprised of the fact that a trip is made between some two cities in the network. It is known from the theory of information that if all probabilities $p_{ij} (i=1,m;\ j=1,n)$ are equivalent, then entropy is maximum. In other words, by finding the maximum function defined by equation (1.36), we would calculate the probabilities of trips being made between all cities in the network under the assumption that all cities in this network generate or attract the same number of trips. It is clear that this assumption is not correct since different cities generate or attract different numbers of trips. So when calculating probabilities p_{ij} some other variables must also be introduced that take greater consideration of the number of trips that are generated or attracted by different cities.

If we use the assumption that all probabilities are equivalent, we can calculate them by solving the following problem :

$$H = - \sum_{i=1}^{m} \sum_{j=1}^{n} p_{ij} \ln p_{ij} \rightarrow max \qquad (1.37)$$

$$\sum_{i=1}^{m} \sum_{j=1}^{n} p_{ij} = 1 \qquad (1.38)$$

where $0 \leqslant p_{ij} \leqslant 1$.

By introducing Lagrangian function F and the Lagrange multiplier λ , we get :

$$F = - \sum_{i=1}^{m} \sum_{j=1}^{n} p_{ij} \ln p_{ij} - \lambda \left(\sum_{i=1}^{m} \sum_{j=1}^{n} p_{ij} - 1 \right) \rightarrow max \quad (1.39)$$

By finding the partial derivations and setting them equal to zero, we get :

$$\frac{\partial F}{\partial p_{ij}} = - \ln p_{ij} - 1 - \lambda = 0 \quad (1.40)$$

$$\frac{\partial F}{\partial \lambda} = - \sum_{i=1}^{m} \sum_{j=1}^{n} p_{ij} + 1 = 0 \quad (1.41)$$

or :

$$p_{ij} = \frac{1}{m \cdot n} \quad (1.42)$$

where entropy is maximal and equals :

$$H_{max} = \ln n\, m \quad (1.43)$$

The proportional model is a special case of the entropy model that has built-in restrictions concerning the number of trips that are generated or attracted by individual regions. In this case probabilities p_{ij} are obtained by solving the following problems :

$$H = - \sum_{i=1}^{m} \sum_{j=1}^{n} p_{ij} \ln p_{ij} \quad (1.44)$$

$$\sum_{j=1}^{n} p_{ij} = u_i \qquad \forall_i \quad (1.45)$$

$$\sum_{i=1}^{m} p_{ij} = v_j \qquad \forall_j \quad (1.46)$$

$$\sum_{i=1}^{m} \sum_{j=1}^{n} p_{ij} = 1 \quad (1.47)$$

where $0 \leq p_{ij} \leq 1$.

Lagrangian function F in this case reads :

$$F = -\sum_{i=1}^{m}\sum_{j=1}^{n} p_{ij} \ln p_{ij} - \sum_{i=1}^{m} \lambda_i \left(\sum_{j=1}^{n} p_{ij} - u_i \right) \tag{1.48}$$

$$- \sum_{j=1}^{n} \left(\mu_j \left(\sum_{i=1}^{m} p_{ij} - \nu_j \right) - \Theta \left(\sum_{i=1}^{m}\sum_{j=1}^{n} p_{ij} - 1 \right) \right)$$

By finding the partial derivations and setting them equal to zero, we get :

$$\frac{\partial F}{\partial p_{ij}} = - \ln p_{ij} - 1 - \lambda_i - \mu_j - \Theta = 0 \tag{1.49}$$

$$\frac{\partial F}{\partial \lambda_i} = - \sum_{j=1}^{n} p_{ij} + u_i = 0 \tag{1.50}$$

$$\frac{\partial F}{\partial \mu_j} = - \sum_{i=1}^{m} p_{ij} + \nu_j = 0 \tag{1.51}$$

$$\frac{\partial F}{\partial \Theta} = - \sum_{i=1}^{m}\sum_{j=1}^{n} p_{ij} + 1 = 0 \tag{1.52}$$

After equations (1.49)-(1.52) have been solved, we get:

$$p_{ij} = u_i \, \nu_j \tag{1.53}$$

or :

$$f_{ij} = \frac{a_i \cdot b_j}{\nu} \tag{1.54}$$

Based on equation (1.53) it can be concluded that the probability of a trip arising between zone i and zone j equuls the product of the probability of a trip being generated in zone i, u_i and the probability of a trip being attracted by zone j, v_j. Equation (1.54) is obtained on the basis of equation (1.53).

As can be concluded from equation (1.54), the number of trips between region i and region j is proportional to the number of trips generated by region i and the number of trips attracted by region j. The proportional model does not include a single factor that creates trip "resistance".

Teodorovic et all [56] used the proportional model to estimate the number of trips between the industrial cities of the continental part of Yugoslavia and the tourist cities on the Adriatic Coast (Figure 1.5).

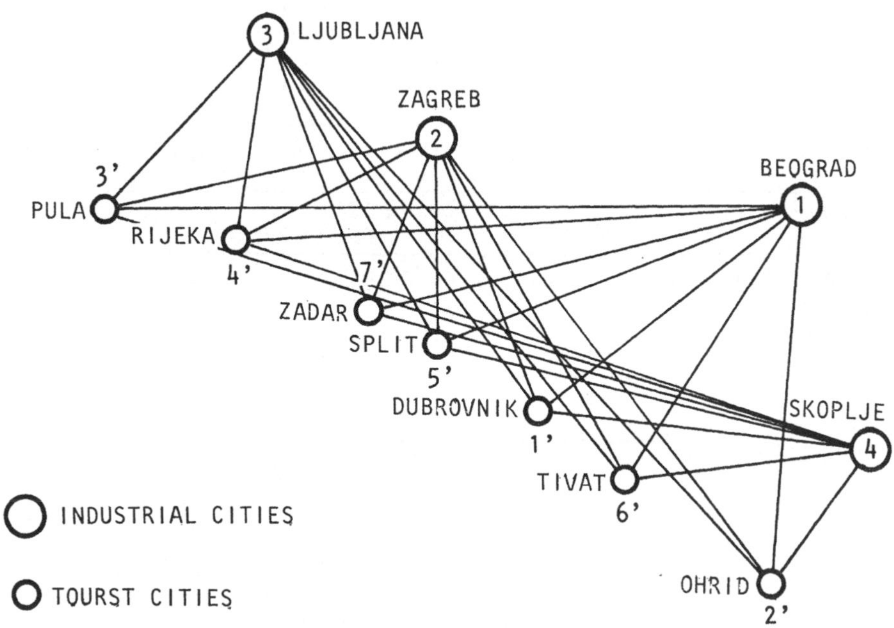

FIGURE 1.5 Route network in Yugoslav domestic air transportation where flows are estimated by the proportional model

Passenger flows were estimated from the industrial cities (Belgrade, Zagreb, Ljubljana, Skoplje) towards the tourist cities (Dubrovnik, Ohrid, Pula, Rijeka, Split, Tivat, Zadar). The number of passengers departing for tourist airports and the number of passengers arriving at tourist airports during 1979 are shown on table 1.4.

TABLE 1.4 **Number of air passengers between industrial and tourist cities in Yugoslavia during 1979**

City i	*No.of passengers from city i departing to tourist airports in 1979.* a_i	*City* j	*No.of passengers arriving in city j from industrial cities in 1979* b_j
Beograd (1)	*362.504*	*Dubrovnik (1)*	*216.831*
Zagreb (2)	*260.293*	*Ohrid (2)*	*14.611*
Ljubljana (3)	*13.822*	*Pula (3)*	*58.234*
Skoplje (4)	*18.837*	*Rijeka (4)*	*26.322*
	$\nu = 655.456$	*Split (5)*	*243.310*
		Tivat (6)	*43.733*
		Zadar (7)	*52.415*
			$\nu = 655.456$

The number of passengers between cities i and cities j is defined as :

$$f_{ij} = \frac{a_i \cdot b_j}{\nu} \tag{1.55}$$

Table 1.5 presents the actual and estimated number of passengers between Belgrade and different tourist cities.

TABLE 1.5 Actual and estimated number of air passengers between Belgrade and tourist cities in Yugoslavia during 1979

Pairs of cities	*Actual No.of passengers in 1979.*	*Estimated No. of passengers in 1979*
Beograd-Dubrovnik	*105.167*	*119.689*
Beograd-Ohrid	*14.611*	*7.976*
Beograd-Pula	*37.948*	*31.900*
Beograd-Rijeka	*26.221*	*14.500*
Beograd-Split	*114.528*	*135.213*
Beograd-Tivat	*43.733*	*23.925*
Beograd-Zadar	*20.296*	*28.638*

In order to estimate the number of air passengers between the industrial and tourist cities in Yugoslavia, Teododrovic et al also used a somewhat modified proportional model where the probability of some trip p_i being made is defined as follows :

$$p_i = \frac{(\text{No.of trips to tourist cities from city } i)}{(\text{Population of city } i)} = \frac{a_i}{S_i} \tag{1.56}$$

The probability of attracting trips p_j is defined as:

$$p_j = \frac{\left(\begin{array}{l}\textit{No. of passengers arriving in} \\ \textit{city j from industrialized} \\ \textit{cities during one year}\end{array}\right)}{\left(\begin{array}{l}\textit{No. of passengers departing} \\ \textit{from industrial cities to all} \\ \textit{tourist airports during one year}\end{array}\right)} = \frac{b_j}{\sum_{j=1}^{n} b_j} \qquad (1.57)$$

where n is the number of tourist cities.

The probability of a trip being made between city i and city j equals :

$$p_{ij} = p_i \cdot p_j = \frac{a_i}{S_i} \cdot \frac{b_j}{\sum_{j=1}^{n} b_j} \qquad (1.58)$$

The total number of passengers traveling from city i to city j is :

$$f_{ij} = p_i \cdot p_j \cdot S_i = \frac{a_i \cdot b_j}{\sum_{j=1}^{n} b_j} \qquad (1.59)$$

The real and estimated number of passengers between Belgrade and the tourist cities obtained from equation (1.59) are shown in Table 1.6 for the period 1971-1979.

As already mentioned, the proportional model does not encompass any factors that create trip "resistance". The proportional model will better reflect reality if it includes some of these factors. We denote by c_{ij} the "cost" of traveling between region i and region j . Cost can

TABLE 1.6 Comparison between actual and estimated number of air passengers between Belgrade and tourist cities in Yugoslavia from 1971 - 1979.

Route	1971		1972.		1973.		1974.		1975.		1976.		1977.		1978.		1979.	
	ADTUAL NO. OF PASSENGERS (a)	ESTIMATED NO. OF PASSENGERS (b)	ADTUAL NO OF PASSENGERS (a)	ESTIMATED NO OF PASSENGERS (b)	ADTUAL NO OF PASSENGERS (a)	ESTIMATED NO OF PASSENGERS (b)	ADTUAL NO OF PASENGERS (a)	ESTIMATED NO OF PASENGERS (b)	ADTUAL NO OF PASENGERS (a)	ESTIMATED NO OF PASSENGERS (b)	ADTUAL NO OF PASSENGERS (a)	ESTIMATED NO OF PASSENGERS (b)	ADTUAL NO OF PASSENGERS (a)	ESTIMATED NO OF PASSENGERS (b)	ADTUAL NO OF PASSENGERS (a)	ESTIMATED NO OF PASENGERS (b)	ADTUAL NO OF PASSENGERS (a)	ESTIMATED NO OF PASSENGERS (b)
BEG-DBV	42222	37580	44260	47110	53717	61670	63462	75710	65251	87260	72408	90390	78814	96900	90233	108180	105167	119980
BEG-OHD	3710	2040	4059	2560	3884	3350	5062	4110	8119	4740	11432	4910	9325	5260	12104	5880	14611	6520
BEG-PUY	2631	7040	7071	8820	11141	11550	17729	14180	21243	16340	22513	16930	20450	18150	31173	20260	37948	22470
BEG- RJK	8603	4770	9589	5980	11357	7820	14225	9600	18201	11070	20985	11470	21603	12290	24331	13720	26221	15220
BEG-SPU	39793	42910	42657	53800	62338	70420	73836	86460	91323	99640	87944	103230	91898	110660	102991	123540	114528	137020
BEG-TIV	8623	9190	24348	11530	30868	15090	39163	18250	42161	21350	38933	22120	53538	23710	46582	26470	43733	29360
BEG-ZAD	7951	9990	10359	12520	13012	16390	15262	20120	17355	23200	18789	24030	17144	25760	19413	28760	20296	31900

include the price of the trip, travel time, a certain combination of these values or some other value. Total "costs" along the entire network equal :

$$C = \sum_{i=1}^{m} \sum_{j=1}^{n} f_{ij} \, c_{ij} \tag{1.60}$$

Average "costs" per passenger equal :

$$c = \sum_{i=1}^{m} \sum_{j=1}^{n} p_{ij} \, c_{ij} \tag{1.61}$$

"Cost" often comprises trip length as well, and the proportional model with a built-in restriction (1.61) is known as the mean trip length model. In this model, probability p_{ij} is obtained by solving the following problem :

$$H = - \sum_{i=1}^{m} \sum_{j=1}^{n} p_{ij} \ln p_{ij} \rightarrow max \tag{1.62}$$

$$\sum_{j=1}^{n} p_{ij} = u_i \tag{1.63}$$

$$\sum_{i=1}^{m} p_{ij} = v_j \tag{1.64}$$

$$\sum_{i=1}^{m} \sum_{j=1}^{n} p_{ij} = 1 \tag{1.65}$$

$$\sum_{i=1}^{m} \sum_{j=1}^{n} p_{ij} \, c_{ij} = c \tag{1.66}$$

Sasaki[137] and Tomlin and Tomlin[165] developed the mean trip length model and tested it on specific numerical examples. Probabilities p_{ij} cannot be expressed as explicitly in the mean trip length model as they can in the proportional model. When introducing the Lagrange multiplier, the Lagrangian function reads :

$$F = - \sum_{i=1}^{m} \sum_{j=1}^{n} p_{ij} \ln p_{ij} - \sum_{i=1}^{m} \lambda_i (\sum_{j=1}^{n} p_{ij} - u_i) -$$

$$- \sum_{j=1}^{n} \mu_j (\sum_{i=1}^{m} p_{ij} - \nu_j) - \Theta (\sum_{i=1}^{m} \sum_{j=1}^{n} p_{ij} - 1) -$$

$$- \rho (\sum_{i=1}^{m} \sum_{j=1}^{n} p_{ij} c_{ij} - c) \qquad (1.67)$$

By finding partial derivations and setting them equal to zero, we get :

$$p_{ij} = e^{-1-\Theta-\lambda_i-\mu_j-c_{ij}} \qquad (1.68)$$

$$u_i = e^{-1-\Theta-\lambda_i} \sum_{j=1}^{n} e^{-\mu_j-\rho c_{ij}} \qquad (1.69)$$

$$\nu_j = e^{-1-\Theta-\mu_j} \sum_{i=1}^{m} e^{-\lambda_i-\rho c_{ij}} \qquad (1.70)$$

$$1 = e^{-1-\Theta} \sum_{i=1}^{m} \sum_{j=1}^{n} e^{-\lambda_i-\mu_j-\rho c_{ij}} \qquad (1.71)$$

$$c = e^{-1-\Theta} \sum_{i=1}^{m} \sum_{j=1}^{n} c_{ij} e^{-\lambda_i-\mu_j-\rho c_{ij}} \qquad (1.72)$$

Using substitutions :

$$q_{ij} = e^{-\rho c_{ij}} \tag{1.73}$$

$$k_i = e^{-1-\Theta-\lambda_i} \tag{1.74}$$

$$l_j = e^{-\mu_j} \tag{1.75}$$

this system of equations is obtained :

$$\sum_{j=1}^{n} k_i\, g_{ij}\, l_j = u_i \tag{1.76}$$

$$\sum_{i=1}^{m} k_i\, g_{ij}\, l_j = \nu_j \tag{1.77}$$

$$\sum_{i=1}^{m} \sum_{j=1}^{n} c_{ij}\, k_i\, g_{ij}\, l_j = c \tag{1.78}$$

The system of equations (1.76) - (1.78) can be solved by the iterative process. Potts and Oliver [120] recommend that starting value $\frac{1}{c}$ be taken first of all for variable ρ . The values of l_j are also chosen arbitrarily in the first step. It is most often taken that $l_j^{(1)} = \nu_j$, where $l_j^{(1)}$ denotes the values of y_j in the first iteration step. We now have :

$$k_i^{(1)} = \frac{u_i}{\sum_{j=1}^{n} q_{ij}\, l_j^{(1)}} \tag{1.79}$$

The values of l_j in the second iteration step are :

$$l_j^{(2)} = \frac{\nu_j}{\sum_{i=1}^{m} k_i^{(1)}\, q_{ij}} \tag{1.80}$$

The values of k_i and l_j in the k-th iteration step are :

$$l_j^{(k)} = \frac{\nu_j}{\sum_{i=1}^{m} k_i^{(k-1)} \cdot q_{ij}} \tag{1.81}$$

$$k_i^{(k)} = \frac{u_j}{\sum_{j=1}^{n} q_{ij} \cdot l_j^{(k)}} \tag{1.82}$$

Probability p_{ij} after k iteration steps is :

$$p_{ij}^{(k)} = k_i^{(k)} \; q_{ij} \; l_j^{(k)} \tag{1.83}$$

The iterative process is finished when :

$$\max_{i,j} \{ | p_{ij}^{(k)} - p_{ij}^{(k-1)} | \} \leqslant \delta \tag{1.84}$$

where δ is the value given in advance.

If equation (1.84) is satisfied, confirmation must be made that the following is satisfied :

$$\sum_{i=1}^{m} \sum_{j=1}^{n} c_{ij} \; k_i \; q_{ij} \; l_j = c \tag{1.85}$$

To this effect, we arbitrarily chose $\frac{1}{c}$ for variable ρ at the beginning, where the value of c was also chosen arbitrarily, so it must be confirmed that the newly obtained value for c equals the earlier arbitrarily chosen value. If this is not the case, i.e. if equation (1.85) is not satisfied, the entire process is repeated with a new value for c, and therefore ρ .

Teodorovic and Acimovic-Raspopovic[154] applied the entropic mean trip length model to estimate passenger flows in Yugoslav domestic air traffic. They took "costs" or the trip "length" to be the ratio between air fare and the time saving achieved by taking the airplane instead of some other means of transportation, so that :

$$c_{ij} = \frac{c'_{ij}}{\Delta t_{ij}} \qquad (1.86)$$

where

c'_{ij} - air fare from city i to city j ,

Δt_{ij} - the time saved by flying from city i to city j instead of driving by car (since in terms of travel time the car is most competitive to the airplane).

Travel time by airplane from city i to city j comprises the time it takes for a passenger to leave the center of city i and arrive in the center of city j . Table 1.7 compares the actual and estimated number of passengers on different routes in Yugoslav domestic air traffic.

The primary merit of the mean trip length model lies in the fact that it needs a relatively small amount of data to be applied. The amount of passenger traffic at different airports is its basic input data. Trip "length" is the corrective factor that has the task of "increasing" the number of passengers on routes with greater time savings and on routes with lower air fare.

TABLE 1.7 Comparison between the actual and estimated number of air passengers on different routes in Yugoslav domestic air traffic in 1977

Route	*Average daily No.of passengers*	*Estimated average daily No. of passengers using mean trip lenght model*
Ljubljana-Beograd	*197*	*138*
Dubrovnik-Beograd	*184*	*240*
Sarajevo-Beograd	*90*	*71*
Rijeka-Beograd	*45*	*56*
Priština-Beograd	*60*	*58*
Sarajevo-Titograd	*10*	*20*
Ljubljana-Dubrovnik	*24*	*33*
Rijeka-Dubrovnik	*6*	*12*
Zagreb-Dubrovnik	*175*	*109*
Ohrid-Beograd	*21*	*21*

1.6.2. The gravity trip distribution model. The gravity trip distribution model originated as an analogy to Newton's law of gravity. According to this law, force F_{ij} between two bodies of mass m_i and m_j located at distance d_{ij} equals:

$$F_{ij} = c \frac{m_i \cdot m_j}{d_{ij}^2} \tag{1.87}$$

where c is constant.

Analogous to this, the number of trips f_{ij} between city i and city j equals :

$$f_{ij} = k \frac{A_i \cdot B_j}{d_{ij}^2} \tag{1.88}$$

where :

A_i - the "size" of city i,

B_j - the "size" of city j,

k - a constant.

City "size" can be taken as the population, number employed, number of trips emitted and received, etc. The essence of gravity models is the assumption that individual cities "attract", thereby creating trips between them in the same way that the mass of certain bodies attracts. The "size" of cities A_i and B_j is most often taken as the number of emitted or attracted trips, i.e. $A_i=a_i$ and $B_j=b_j$ so that :

$$f_{ij} = k \frac{a_i \cdot b_j}{d_{ij}^2} \tag{1.89}$$

Based on equation (1.89) it can be concluded that the number of trips from city i to city j is directly proportional to the total number of trips generated by city i and the total number of trips attracted by city j , and indirectly proportional to the square of the distance between the cities. The main drawback of the model as defined by equation (1.89) is that, for example, the number of trips will increase 9 times if a_i increases 3 times and b_j increases 3 times. The question also arises as to why the distance is squared in the model. In other words, equations :

$$\sum_{j=1}^{n} f_{ij} = a_i \qquad \sum_{i=1}^{m} f_{ij} = b_j$$

are not satisfied by the model as defined by equation (1.89).

In order to satisfy these equations, later modifications of the gravity model introduce coefficients k_i and k_j associated with the number of trips emitted or attracted by the cities. Later modifications also replace squared distance d_{ij}^2 by distance function $f(d_{ij})$. The general form of the modified gravity model reads :

$$f_{ij} = k_i\ a_i\ k_j\ b_j\ f(d_{ij}) \tag{1.90}$$

Distance can mean distance, travel time, a combination of these variables or some other variable.

Since :

$$\sum_{j=1}^{n} f_{ij} = a_i \tag{1.91}$$

then :

$$\sum_{j=1}^{n} k_i\ a_i\ k_j\ b_j\ f(d_{ij}) = a_i \tag{1.92}$$

and :

$$k_i = \frac{1}{\sum_{j=1}^{n} k_j \, b_j \, f(d_{ij})} \tag{1.93}$$

We also have :

$$\sum_{i=1}^{m} f_{ij} = b_j$$

$$\sum_{i=1}^{m} k_i \, a_i \, k_j \, b_j \, f(d_{ij}) = b_j$$

and :

$$k_j = \frac{1}{\sum_{i=1}^{m} k_i \, a_i \, f(d_{ij})} \tag{1.94}$$

Parameters k_i and k_j are defined by equations (1.93) and (1.94), They are calculated using the iterative procedure.

The following form of the gravity model is also often used :

$$f_{ij} = k \, a_i \, b_j \, e^{-cd_{ij}} \tag{1.95}$$

where k and c are parameters that are estimated statistically.

In this case as well, the number of trips f_{ij} from city i to city j increases with a rise in the total number of trips generated by city i, a_i and a rise in the total number of trips attracted by city j, b_j. However, due to the existence of corrective factor k, the number of trips f_{ij} in this case will not increase, for example, four times if a_i and b_j increase twice. We can take variable d_{ij} as before as the price of transportation or some other variable that creates trip "resistance". In order to estimate parameters k and c, and therefore calibrate the model, the model must first be put into linear form. By finding the logarithm of the left and right sides of equation (1.95) we get :

$$ln\ f_{ij} = ln\ k + ln\ a_i\ b_j - c\ d_{ij} \qquad (1.96)$$

and

$$ln\ \frac{f_{ij}}{a_i\ b_j} = ln\ k - c\ d_{ij} \qquad (1.97)$$

By introducing substitutions :

$$F_{ij} = ln \frac{f_{ij}}{a_i\ b_j} \quad , \quad ln\ k = K \quad i \quad -c = b \qquad (1.98)$$

we get :

$$F_{ij} = K + b\ d_{ij} \qquad (1.99)$$

Parameters K and b can be defined by the least squares method. Statistical data must necessarily be collected concerning the n pairs of cities (data on variables f_{ij}, a_i, b_j and d_{ij}). Parameters K and c are defined so that the sum of the deviation squares :

$$S = \sum_{l=1}^{n} (K + b\ d_{ij}^{(l)} - F_{ij}^{(l)})^2 \qquad (1.100)$$

is minimal. (note : $d_{ij}^{(l)}$ denotes the price of transportation between the l-th pair of cities).

By finding the partial derivations and setting them equal to zero, we get :

$$\frac{\partial S}{\partial K} = 2 \sum_{l=1}^{n} (K+bd_{ij}^{(l)} - F_{ij}^{(l)}) = 0 \qquad (1.101)$$

$$\frac{\partial S}{\partial b} = 2 \sum_{l=1}^{n} d_{ij}^{(l)} (K+bd_{ij}^{(l)} - F_{ij}^{(l)}) = 0 \qquad (1.102)$$

Parameters K and b are obtained by solving the system of equations (1.101) - (1.102).

1.7. Multimode Models to Estimate Transportation Demand

The previous presentations underscored that air transportation demand on long-haul routes is studied independently of other modes of transportation since airplanes are the predominant mode of transportation in long-haul traffic. The presence of numerous competitive modes of transportation and a large number of possible routes considerably complicates an evaluation of potential passenger flows. Therefore, air transportation demand on shorter and medium routes is primarily estimated using multimode models that also include the influence of competitive modes of transportation. On shorter and medium routes, the total number of passengers in all modes of transportation between two cities is conditioned above all by the population of the two cities, the number employed and the distance between them. The percentage of passengers that chooses to fly on these routes is influenced by the price and travel time of different modes of transportation, the daily number of departures, the degree that flight and vehicle departure schedules accommodate passenger demand in terms of travel time, the reliability with which the schedule is carried out (cancelling flights and tardiness), travel comfort, etc. Passengers making private trips are particularly sensitive to the price of transportation since they bear the cost themselves. Business passengers are more sensitive to the daily number of departures, schedule accommodation to passenger requests in terms of travel time, comfort and particularly the schedule's reliability. Stojkovic[150] and Kedic[77] surveyed passengers in Yugoslav domestic air traffic and demonstrated that the percentage of business passengers deciding to fly is much greater during the summer period since the weather situation during winter decreases the reliability of the schedule being carried out.

All multimode models can basically be divided into aggregated and disaggregated models. Aggregated models take certain socio-economic characteristics into consideration when estimating the percentage share of air traffic out of the total number of passengers transported. For example, the average national income in a certain city is considered to be one of the more important indicators affecting the number of passengers. Disaggregated models start with the individual as the one making the decision to travel and therefore operate with certain socio-economic characteristics related to the individual, obtained by surveying passengers. Disaggregated models can also quantify the effect of comfort or the feeling of safety that the passenger has when traveling on his decision to use a certain type of transportation. Disaggregated models developed during the past two decades. Although aggregated models are still in use, disaggregated models to a certain extent better describe the process by which a passenger chooses a mode of transportation.

1.8. The Abstract Mode Model to Estimate the Number of Passengers on Intercity Traffic

One of the best-known models to estimate the number of passengers choosing different modes of transportation is the abstract mode model, developed by Quandt and Baumol[123]. The model is applied when elaborating a larger number of transportation studies. Numerous modifications of this model have also been developed. We will herewith present the basic assumptions of the abstract mode model using a suitable numerical example.

Let there be three different modes of transportation between two cities. Costs and travel time per mode are given in Table 1.8.

TABLE 1.8. Travel time and costs per transportation mode

Transportation mode M_i	*Travel costs* c_{Mi}	*Travel time* t_{Mi}
M_1	(1)	10
M_2	2	4
M_3	5	(2)

Transportation mode M_1 has the lowest travel cost. But mode M_3 has the shortest travel time. Relative travel costs or time per transportation mode are defined as the ratio between each mode's travel costs or time and the least possible travel costs or time on the route under observation. In other words, the relative characteristics of transportation modes M_i, $c^r_{M_i}$ $t^r_{M_i}$ in our example are :

$$c^r_{M_i} = \frac{c_{M_i}}{\min_j \{c_{M_j}\}} \quad , \quad j = 1,2,3 \qquad (1.103)$$

$$t^r_{M_i} = \frac{t_{M_i}}{\min_j \{t_{M_j}\}} \quad , \quad j = 1,2,3 \qquad (1.104)$$

A transportation mode's relative characteristics are calculated by dividing each of its characteristics by the best possible value of that characteristic. In the case of travel costs and time, the best value is the least one, while for frequency the best value would be the greatest one.

Relative travel costs and time for the three modes of transportation are given in Table 1.9.

TABLE 1.9 Relative travel costs and time per transportation mode

Transportation mode M_i	Relative travel costs $c^r_{M_i}$	Relative travel time $t^r_{M_i}$
M_1	$c^r_{M_1} = \frac{1}{min\{1,2,5\}} = \frac{1}{1} = 1$	$t^r_{M_1} = \frac{10}{min\{10,4,2\}} = \frac{10}{2} = 5$
M_2	$c^r_{M_2} = \frac{2}{min\{1,2,5\}} = \frac{2}{1} = 2$	$t^r_{M_2} = \frac{4}{min\{10,4,2\}} = \frac{4}{2} = 2$
M_3	$c^r_{M_3} = \frac{5}{min\{1,2,5\}} = \frac{5}{1} = 5$	$t^r_{M_3} = \frac{2}{min\{10,4,2\}} = \frac{2}{2} = 1$

The relative frequencies of the modes of transportation $f^r_{M_i}$ are obtained as the ratio between the frequency of the mode in question and the greatest frequency on the route under observation, i.e. :

$$f^r_{M_i} = \frac{f_{M_i}}{\max_j\{f_{M_j}\}} \quad , \quad j = 1,2,3 \qquad (1.105)$$

The transportation mode frequencies and their relative values are given in Table 1.10.

TABLE 1.10 Relative frequencies

Transportation mode M_i	Frequency (daily No. of departures) f_{M_i}	Relative frequency $f^r_{M_i}$
M_1	6	$f^r_{M_1}=\frac{6}{max\{6,12,4\}}=\frac{6}{12}=0,5$
M_2	12	$f^r_{M_2}=\frac{12}{max\{6,12,4\}}=\frac{12}{12}=1$
M_3	4	$f^r_{M_3}=\frac{4}{max\{6,12,4\}}=\frac{4}{12}=0,33$

Quandt and Baumol started with the assumption that the passenger's choice of transportation mode, or the number of passengers choosing a certain mode, depends on both the best characteristic values of all transportation modes (least costs, least time, greatest frequency) and on the relative characteristic values of the individual mode. The model also has a built in assumption that the total number of trips by all modes of transportation between two cities depends on the socio-economic characteristics of those cities. The original Quandt-Baumol model reads :

$$T_{kij}=\alpha_o P_i^{\alpha 1}\cdot P_j^{\alpha 2}\cdot Y_i^{\alpha 3}\cdot Y_j^{\alpha 4}\cdot M_i^{\alpha 5}\cdot M_j^{\alpha 6}\cdot N_{ij}^{\alpha 7}(H^b_{ij})^{\beta o}(H^r_{kij})^{\beta 1}$$

$$(C^b_{ij})^{\gamma o}(C^r_{kij})^{\gamma 1}(D^b_{ij})^{\delta o}(D^r_{kij})^{\delta 1} \qquad (1.106)$$

where :

T_{kij} - the number of trips between city i and city j by transportation mode k ,

P_i, P_j - the populations in cities i and j ,

Y_i, Y_j - the average national income in cities i and j ,

M_i, M_j - the percentage of the population employed in industry in cities i and j ,

N_{ij} - then number of different transportation modes operating between cities i and j ,

H^b_{ij} - the least possible travel time between cities i and j ,

H^r_{kij} - relative travel time between cities i and j by transportation mode k ,

C^b_{ij} - the least possible travel costs between i and j ,

C^r_{kij} - relative travel costs between i and j by transportation mode k ,

D^b_{ij} - the greatest frequency (number of departures) between cities i and j ,

D^r_{kij} - relative frequency between cities i and j by transportation mode k ,

$\alpha_0, \alpha_1, \alpha_2, \alpha_3, \alpha_4, \alpha_5, \alpha_6, \alpha_7, \beta_0, \beta_1, \gamma_0, \gamma_1, \delta_0, \delta_1$ - parameters estimated statistically.

The number of passengers between city i and city j by transportation mode k , T_{kij} is a dependent variable in the model. The populations of cities i and j , P_i and P_j, the average national incomes Y_i and Y_j, and the percentage of the population employed in industry M_i and M_j, are independent variables that primarily influence the total number of trips generated by all modes of transportation between cities i and j. So the Quandt-Baumol model assumes, as in gravity models, that cities with a greater "mass" generate a greater number of trips between them. The other independent variables figuring in the model affect the number of passengers deciding on specific modes of transportation. The closer the relative characteristics of a specific mode are to the best possible characteristics, the greater the number of passengers choosing this mode of transportation. Model parameters are estimated by applying the multiple regression technique. First the model is put into linear form by finding the logarithm of the left and right sides of equation (1.106), and then the multiple regression technique is applied. Several different variations of the Quandt-Baumol model were tested with actual data when compiling transportation studies in the USA. The model gave good results when applied; it can estimate the number of passengers that would choose a new mode of transportation if introduced on a route. With preliminary model calibration and knowledge of the new mode's characteristics, it would be very easy to estimate the number of passengers that would choose the new mode of transportation.

1.9. Choice Models

Choice models that estimate the percentage share of different modes of transportation out of the total number of trips, or the probability of using a certain mode, start with the individual as the trip decision maker. In addition to the frequency, travel time and cost of different modes of transportation, numerous other factors are present whose effects cannot be quantified without polling the passenger population. Thus, without surveys it is impossible to determine the effects of comfort, the passenger's feeling of safety during the trip or reliability in carrying out the established schedule during summer and winter, on the choice of transportation mode. Sommers [148] and Levin and Herring [91] studied the effects of these nondemographic factors on the choice of flying on routes with competition from different modes of transportation. Polls concerning the importance passengers give to nondemographic factors typically use a scale that enables the passenger to indicate his feelings. For example, the feeling of safety could be determined by the following :

Place an X on the scale below to indicate your feeling of safety when flying from city A and city B .

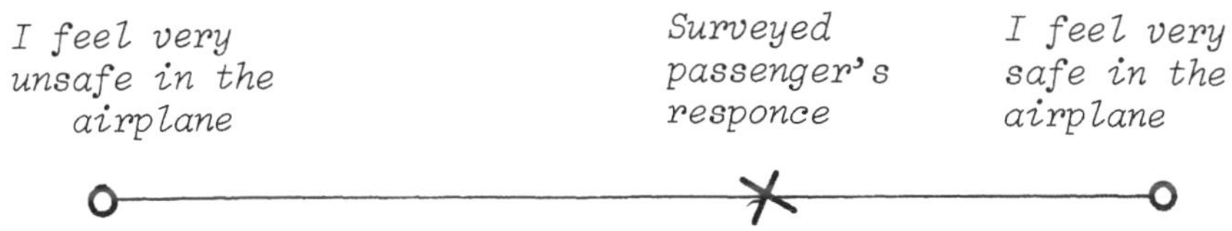

FIGURE 1.6 Safety evaluation scale

If the scale is 10 cm long, for example, a simple measurement of the length from left to right, from the beginning of the scale to the (X) marked by the surveyed passenger, will give a numerical value to the feeling of safety when flying. When taking the survey, a precisely defined travel scenario should be offered to participants. This means that they should indicate whether this is a business trip or a private trip, whether the trip is being made in summer or winter, etc. To this effect, due to the frequent possibility of flights being cancelled or late during the winter, passengers will certainly give a poorer rating to the airplane as a reliable means in terms of executing the schedule on time. In the same vein, the choice of transportation mode vitally depends on the purpose of the trip, making it essential to mention this purpose within the offered travel scenario. The survey questionnaire should therefore contain sentences of this type : Imagine that you are taking a business trip, alone, during the summer...

Sommers used a scale to evaluate nondemographic factors influencing the choice of transportation mode with a view to estimating the future percentage share of V/STOL aircraft out of total transportation on certain routes. The scale can have a somewhat different concept than shown in Figure 1.6. Figure 1.7 shows another possible type of scale to evaluate the feeling of safety.

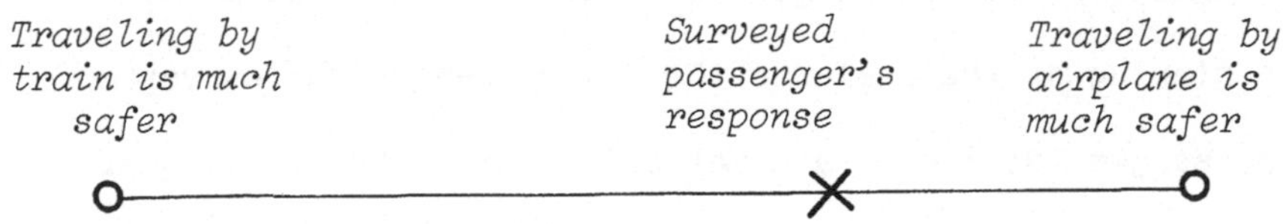

FIGURE 1.7 A possible scale to evaluate the feeling of safety when traveling

The type of scale shown on Figure 1.7 was used by Levin and Herring.

Choice models can be divided into deterministic and stochastic models. Deterministic choice models assume that the process of choosing a mode of transportation is essentially deterministic and that the individual always chooses the mode of transportation in the same manner. In other words, it is assumed that if the individual chose the airplane once when traveling from city A to city B, he will also choose the airplane in subsequent situations. Stochastic choice models assume that the choice process is subject to many random effects that cannot be precisely perceived. As noted by Kanafani[74], stochastic models more realistically describe the process of passenger choice than deterministic models.

Let us note a passenger who is to travel between two cities. We denote by A_i one of the possible alternatives to make this trip $(i=1,2,...,m)$. We denote the number of possible alternatives (modes) by m. To every alternative we join function $V(i)$ called the choice function. The choice function is most often of the form :

$$V(i) = a_1X_{1i} + a_2X_{2i} + ... + a_nX_{ni} \qquad (1.107)$$

where :

$X_{1i}, X_{2i}, ..., X_{ni}$ - variables affecting the choice of mode that relate to alternative A_i. (For example, X_{1i} could be travel time by alternative A_i, X_{2i} travel costs by alternative A_i, etc.),

$a_1, a_2, ..., a_n$ - parameters estimated when calibrating the choice model that indicate the effect of variables $X_{1i}, X_{2i}, ..., X_{ni}$ on the choice of alternative A_i.

Choice models do not have to be presented as a linear function. If choice function $V(i)$ is treated as a utility function that has passengers if alternative A_i is chosen, then the probability of alternative A_i being chosen is greater the greater the value of choice function $V(i)$. It was mentioned earlier that stochastic choice models more realistically describe the process of passenger choice than deterministic models. To this effect, the choice process is subject to many random effects. The reason for this is that passengers give a different evaluation to the utility they have by opting for a specific mode of transportation. This difference in the evaluation of utility results from different evaluations of variables in the choice function. For example, passengers often do not have complete information on the frequency of competitive modes of transportation or on departure schedules. In some cases, passengers do not know the cost of some of the modes. In the same vein, some passengers do not opt for a certain alternative even though it would be very logical to do so because of its advantages compared to other alternatives.

We denote by $U(i)$ the function of estimated utility achieved by opting for alternative A_i. Random variable $U(i)$ is defined as :

$$U(i) = V(i) + e(i) \qquad (1.108)$$

where $e(i)$ is a random variable subject to a certain probability density function and reflects the stochastic nature of the choice of alternative.

In stochastic choice models, the choice function is taken as function $U(i)$. We mention once more that $V(i)$ is a deterministic function of the attributes of alternative A_i. Let us consider the case where $U(i)$ takes on the value $U^{(1)}(i)$, and $e(i)$ the value $e^{(1)}(i)$.

All choice models start with the assumption that the passenger will choose alternative A_i if it corresponds to the greatest value of the choice function. This means that the probability of choosing alternative A_i when $U(i)=U^1(i)$ and $e(i)=e^{(1)}(i)$ equals :

$$p^{(1)}(i) = p[U^{(1)}(i) > U^{(1)}(j), \quad \forall j \neq i] \qquad (1.109)$$

or :

$$p^{(1)}(i)=p[V(i) + e^{(1)}(i)>V(j) + e^{(1)}(j), \quad \forall j\neq i] \qquad (1.110)$$

i.e. :

$$p^{(1)}(i)=p[e^{(1)}(j)<V(i)-V(j) + e^{(1)}(i), \quad \forall j \neq i] \qquad (1.111)$$

We denote by $F[\cdot]$ cumulative density function $e(i), e(j), e(k)$,

Since :

$$p[e^{(1)}(j)<V(i)-V(j)+e^{(1)}(i)]=F[V(i)-V(j)+e^{(1)}(i), \forall j\neq i] \qquad (1.112)$$

then :

$$p^{(1)}(i)=F[V(i)-V(j)+e^{(1)}(i), \quad \forall j\neq i] \qquad (1.113)$$

For some other value of random variable $e(i)$, for example $e(i)=e^{(2)}(i)$, we have :

$$p^{(2)}(i)=F[V(i)-V(j)+e^{(2)}(i), \quad \forall j\neq i] \qquad (1.114)$$

For $e(i)=e^{(3)}(i)$ we have :

$$p^{(3)}(i)=F[V(i)-V(j)+e^{(3)}(i), \quad \forall j \neq i] \qquad (1.115)$$

We denote the probability density function of random variable $e(i)$ by $f[e(i)]$. We can finally write that the probability of choosing alternative A_i equals :

$$p(i)=\int_{e(i)} F[V(i)-V(j)+e(i),\forall j\neq i]\,f[e(i)]\,d[e(i)] \quad (1.116)$$

For different probability density functions of the random variable $e(i)$, different stochastic choice models are obtained.

1.9.1. LOGIT Model. One of the best-known stochastic choice models is the LOGIT model. This model is based on the assumption that random variables $e(i)$, $e(j)$, $e(k)$,...are independent and distributed by Gumbel's probability density function. Gumbel's probability density function is also known as double exponential density. Probability density function $f(x)$ of random variable X distributed according to Gumbel's density equals :

$$f(x)=\lambda e^{-x}\cdot e^{-\lambda e^{-x}}, \quad \lambda>0, \quad -\infty<x<+\infty \quad (1.117)$$

Cumulative density function $F(x)$ equals:

$$F(x) = e^{-\lambda e^{-x}} \quad (1.118)$$

We saw that the probability of choosing alternative A_i equals:

$$p(i) = p[U(i)>U(j), \quad \forall j\neq i] \quad (1.119)$$

Let there be z possible alternatives or possible choices of transportation mode. Equation (1.119) written in another way reads :

$$p(i)=p\left[U(i)>U(1)\right]\cdot p\left[U(i)>U(2)\right]\ldots p\left[U(i)>U(i-1)\right]\cdot$$
$$\cdot p\left[U(i)>U(i+1)\right]\ldots p\left[U(i)>U(Z)\right] \qquad (1.120)$$

or

$$p(i) = \prod_{j\neq i} p\left[U(i)>U(j)\right] \qquad (1.121)$$

Analogous to this, we can write equation (1.116) as :

$$p(i)=\int_{e(i)} \prod_{j\neq i} F\left[V(i)-V(j)+e(i)\right] f\left[e(i)\right] d\left[e(i)\right] \qquad (1.122)$$

In order to simplify the terms we denote random variable $e(i)$ by y. For double exponential density, random variable $e(i)$ will be :

$$F\left[V(i)-V(j)+y\right]= e^{-\lambda e^{-\left[V(i)-V(j)+y\right]}} \qquad (1.123)$$

We also have :

$$f(y) = \lambda e^{-y} \cdot e^{-\lambda e^{-y}} \qquad (1.124)$$

By substituting (1.123) and (1.124) in (1.122) we get :

$$p(i) = \int_{-\infty}^{+\infty} \prod_{j\neq i} e^{-\lambda e^{-\left[V(i)-V(j)+y\right]}} \cdot \lambda e^{-y} e^{-\lambda e^{-y}} \cdot dy \qquad (1.125)$$

Since for $j = i$: $e^{-\lambda e^{-y}} = e^{-\lambda e^{-[V(i)-V(j)+y]}}$, then :

$$p(i) = \int_{-\infty}^{+\infty} \prod_j e^{-\lambda e^{-[V(i)-V(j)+y]}} \cdot \lambda e^{-y} dy \qquad (1.126)$$

or :

$$p(i) = \int_{-\infty}^{+\infty} e^{-\lambda e^{-y} \cdot \sum_j e^{V(j)-V(i)}} \cdot \lambda e^{-y}\, dy \qquad (1.127)$$

By introducing the substitution :

$$t = -\lambda e^{-y} \sum_j e^{V(j)-V(i)}$$

and solving the integral, we get :

$$p(i) = \frac{1}{\sum_j e^{V(j)-V(i)}} \cdot e^{-\lambda e^{-y} \sum_j e^{V(j)-V(i)}} \Big|_{-\infty}^{+\infty} \qquad (1.128)$$

Since :

$$e^{-\lambda e^{-y} \sum_j e^{V(j)-V(i)}} \Big|_{-\infty}^{+\infty} = 1 \qquad (1.129)$$

then :

$$p(i) = \frac{1}{\sum_j e^{V(j)-V(i)}} \qquad (1.130)$$

and :

$$p(i) = \frac{e^{V(i)}}{\sum_j e^{V(j)}} \qquad (1.131)$$

In order to calculate choice probabilities $p(i)$

for different alternatives A_i, when defining choice function $V(i)$ an estimate must be made of the parameters figuring in the function. The choice function is often not linear so the least squares method or the multiple regression technique can only be used to estimate the parameters of the choice function in special cases. The maximum likelihood method is used to estimate the parameters of choice functions. Choice models can be based on disaggregated or aggregated data. Choice models based on disaggregated data presume that each passenger evaluates the advantages and defects of each alternative differently, so that the passenger population must be polled before estimating the parameters of the choice function. In this case each polled passenger corresponds to a different choice function value. For choice models based on aggregated data, all passengers correspond to the same choice function value since it is assumed that the variables in the choice function are equal for all passengers. Therefore, the probability of choosing a certain alternative for aggregated data is equal for all passengers.

Choice models that only have two alternatives (two transportation mode alternatives) are called binomial choice models. For three, four or more possible alternatives, these are called multinomial choice models.

1.9.2. <u>Calibrating the LOGIT model</u>. To illustrate the maximum likelihood method, we will estimate the parameters of a choice function for a multinomial LOGIT model based on disaggregated data.

Let there be a total of n polled passengers and let there be a total of z possible alternatives (modes of transportation). We note the r th passenger and the s th alternative. The choice function of the s-th alternative made by the r th passengers is denoted by $V(r,s)$. We also assume that function $V(r,s)$ is linear, i.e. :

$$V(r,s) = a_{1s}X_{1rs}+a_{2s}X_{2rs}+ \dots +a_{ms}X_{mrs} \quad , \; r=1,2,\dots,n \; , \; s=1,2,\dots,z \tag{1.132}$$

where :

m - the number of variables in the choice function,

$X_{1rs}, X_{2rs}, \dots, X_{mrs}$ - the value of the variables in the choice function evaluated by the r th passengers when estimating the s -th alternative.

We denote by p_{rs} the probability that the s -th alternative will be chosen by the r th passenger. For the LOGIT model, this will be :

$$p_{r,s} = \frac{e^{V(r,s)}}{\sum_{t=1}^{z} e^{V(r,t)}} = \frac{e^{a_{1s}X_{1rs}+a_{2s}X_{2rs}+\dots+a_{ms}X_{mrs}}}{\sum_{t=1}^{z} e^{a_{1t}X_{1rt}+a_{2t}X_{2rt}+\dots+a_{mt}X_{mrt}}} \tag{1.133}$$

During polling, passengers decide on a specific mode of transportation in addition to allotting values to the choice function's variables. Out of n polled passengers, let there be n_1 who choose alternative A_1 , n_2 passengers who choose alternative A_2,... n_z passengers for alternative A_z
It is clear that :

$$n_1 + n_2 + \dots + n_z = n \qquad (1.134)$$

The set of passengers that choose alternative A_1 is denoted by Q_1 , the set of passengers choosing A_2 is denoted by Q_2 , etc.

We now calculate the probability that out of n polled passengers, n_1 choose alternative A_1, n_2 choose alternative A_2,... and n_z passengers choose alternative A_z . One of the possibilities for such an event taking place is that the first n_1 polled passengers choose alternative A_1 , the following n_2 passengers choose alternative A_2, etc. The probability of this happening is :

$$P = (p_{1,1} \cdot p_{2,1} \cdots p_{n_1,1}) \cdot (p_{n_1+1,2} \cdot p_{n_1+2,2} \cdots p_{n_1+n_2,2}) \cdots (p_{\sum_{i=1}^{z-1} n_i+1,z} \cdots p_{n,z}) \qquad (1.135)$$

or :

$$P = \prod_{s=1}^{z} \prod_{r \in Q_s} p_{rs} \qquad (1.136)$$

However, probability P is the probability of achieving one of the possibilities that of n polled passengers, n_1 pick alternative A_1, n_2 alternative A_2 ,... and n_z

alternative A_z. We then calculate the total number of such possibilities. Out of n passengers, n_1 choose alternative A_1. This choice can be made in $\binom{n}{n_1}$ different ways. Of the remaining $(n-n_1)$ passengers, n_2 will choose another mode of transportation. This choice can be made in $\binom{n-n_1}{n_2}$ different ways. Choices in the first two modes of transportation can be made in

$$\binom{n}{n_1} \cdot \binom{n-n_1}{n_2} \text{ different ways.}$$

We designate that :

$$\binom{n}{n_1} \cdot \binom{n-n_1}{n_2} = \frac{n!}{n_1!n_2!} \cdot \frac{1}{(n-n_1-n_2)!} \qquad (1.137)$$

Analogous to the above, we conclude that choosing the first three modes of transportation can be made in :

$$\frac{n!}{n_1!n_2!n_3!} \cdot \frac{1}{(n-n_1-n_2-n_3)!} \qquad \text{different ways}$$

Finally, we can write that the total number of ways in which n_1 polled passengers out of n can choose alternative A_1, n_2 can choose alternative $A_2, \ldots,$ and n_z can choose alternative A_z equals :

$$\frac{n!}{n_1!n_2!\ldots n_z!} \cdot \frac{1}{(n-n_1-n_2-\ldots-n_z)!} = \frac{n!}{n_1!n_2!\ldots n_z!} \qquad (1.138$$

Probability $P(n_1,n_2,\ldots,n_z)$ that n_1 passengers out of polled will choose alternative $A_1, \ldots,$ and n_z will choose alternative A_z equals :

$$P(n_1,n_2,\ldots,n_z)=\frac{n!}{n_1!n_2!\ldots n_z!}\prod_{s=1}^{z}\prod_{r\in Q_s} p_{rs} \qquad (1.139)$$

We designate the probability density defined by formula (1.139) to be a polynomial distribution. Values $n_1,n_2,\ldots n_z$ are the results of experiments concerning n passengers' choice of mode. Likelihood function $L(n_1,n_2,\ldots,n_z;\ W_1,W_2,\ldots,W_q)$ comprises the probability that the sample of scope n is composed of components $n_1,n_2,\ldots,n_z$, where $W_1,W_2,\ldots W_q$ are parameters that figure in the probability density function. In our case the likelihood function is exactly the same as the probability defined by equation (1.139) :

$$L = \frac{n!}{n_1!n_2!\ldots n_z!}\prod_{s=1}^{z}\prod_{r\in Q_s} p_{rs} \qquad (1.140)$$

The estimate of maximum likelihood $\hat{W}_1,\hat{W}_2,\ldots,\hat{W}_q$ of parameters $W_1,W_2,\ldots,W_q$ for the given sample $n_1,n_2,\ldots,n_z$ are those parameters where likelihood function L reaches a maximum.

By finding the logarithm of the left and right sides of equation (1.140), we get :

$$lnL = ln\frac{n!}{n_1!n_2!\ldots n_z!} + \sum_{s=1}^{z}\sum_{r\in Q_s} ln\ p_{rs} \qquad (1.141)$$

We denote the following expression by $\bar{L}$:

$$\bar{L} = \sum_{s=1}^{z}\sum_{r\in Q_s} ln\ p_{rs} = \sum_{s=1}^{z}\sum_{r\in Q_s} ln\frac{e^{V(r,s)}}{\sum_{t=1}^{z} e^{V(r,t)}} \qquad (1.142)$$

$$\bar{L} = \sum_{s=1}^{z} \sum_{r \in Q_s} [V(r,s) - ln \sum_{t=1}^{z} e^{V(r,t)}] \quad (1.143)$$

$$\bar{L} = \sum_{s=1}^{z} \sum_{r \in Q_s} [\sum_{i=1}^{m} a_{is} X_{irs} - ln \sum_{t=1}^{z} e^{\sum_{i=1}^{m} a_{is} X_{irs}}]$$

In our case a_{is} $(i=1,2,\ldots,m;\ s=1,2,\ldots,z)$ are the parameters that we must estimate. We arrive at these estimates by finding the corresponding partial derivations and setting them equal to zero :

$$\frac{\partial \bar{L}}{\partial a_{is}} = 0 \ , \quad s = 1,2,\ldots,z \quad (1.145)$$

$$\frac{\partial \bar{L}}{\partial a_{is}} = \sum_{s=1}^{z} \sum_{r \in Q_s} X_{irs} - \sum_{s=1}^{z} \sum_{r \in Q_s} X_{irs} \cdot \frac{e^{\sum_{i=1}^{m} a_{is} X_{irs}}}{\sum_{t=1}^{z} e^{\sum_{i=1}^{m} a_{is} X_{irs}}} = 0$$

$$s = 1,2,\ldots,z \quad (1.146)$$

Equation (1.146) can be solved by iterative numerical procedures. We will illustrate how to determine the values of parameters a_{is} on the following simple numerical example that concerns the choice of one out of two possible modes of transportation (a binomial disaggregated model).

We assume that businessmen on a certain route traveling during the winter choose between the airplane and train uniquely on the basis of the departure schedule's reliability. Let there be 7 passengers who have decided on a certain mode of transportation and who have evaluated the reliability of the airplane and the train in terms of carrying out their departure schedules. Let there be a possible evaluation of 0 to 10 with 0 denoting a totally unreliable mode of transportation, and 10 an exceptionally dependable mode of transportation in terms of departures. Table 1.11 presents each passenger's choice of transportation mode and his estimated reliability of the departure schedule. We denote the choice of airplane by A_1 and the choice of train by A_2.

The first three polled passengers chose the airplane ($Q_1=\{1,2,3\}$), and the fourth, fifth, sixth and seventh passengers chose the train ($Q_2=\{4,5,6,7\}$).

TABLE 1.11 Passenger Choice of Transportation Mode

Polled passenger r	*Mode choice* A_i	*Estimated airplane departure schedule reliability* X_{1r1}	*Estimated train deperture schedule reliability* X_{1r2}
1	A_1	9	7
2	A_1	8	6
3	A_1	9	8
4	A_2	7	10
5	A_2	8	9
6	A_2	5	7
7	A_2	9	10

We assume that this is a very simple choice function :

$$V(r,s) = a_{1s} X_{1rs} \tag{1.147}$$

where a_{1s} are parameters to be estimated, and X_{1rs} the r-th passenger's evaluation of the departure schedule reliability of the s-th mode of transportation $(s=1,2)$.

Probability p_{rs} that the r-th passenger will choose the s-th mode of transportation equals :

$$p_{rs} = \frac{e^{a_{1s}X_{1rs}}}{\sum_{s=1}^{2} e^{a_{1s}X_{1rs}}} \tag{1.148}$$

i.e. the probability of the r-th passenger choosing the airplane is :

$$p_{r1} = \frac{e^{a_{11}X_{1r1}}}{e^{a_{11}X_{1r1}} + e^{a_{12}X_{1r2}}} \tag{1.149}$$

and the probability of the r-th passenger choosing the train is :

$$p_{r2} = \frac{e^{a_{12}X_{1r2}}}{e^{a_{11}X_{1r1}} + e^{a_{12}X_{1r2}}} \tag{1.150}$$

In our case the likelihood function reads:

$$L = \frac{7!}{3!4!} \cdot (p_{11} \cdot p_{21} \cdot p_{31}) \cdot (p_{42} \cdot p_{52} \cdot p_{62} \cdot p_{72}) \tag{1.151}$$

so that :

$$\ln L = \ln \frac{7!}{3!4!} + (\ln p_{11} + \ln p_{21} + \ln p_{31}) + (\ln p_{42} + \ln p_{52} + \ln p_{62} + \ln p_{72}) \qquad (1.152)$$

or :

$$\bar{L} = \sum_{r \in Q_1} \ln p_{r1} + \sum_{r \in Q_2} \ln p_{r2} \qquad (1.153)$$

i.e. :

$$\bar{L} = \sum_{r \in Q_1} \ln\left(\frac{e^{a_{11}X_{1r1}}}{e^{a_{11}X_{1r1}} + e^{a_{12}X_{1r2}}}\right) + \sum_{r \in Q_2} \ln\left(\frac{e^{a_{12}X_{1r2}}}{e^{a_{11}X_{1r1}} + e^{a_{12}X_{1r2}}}\right) \qquad (1.154)$$

$$\bar{L} = \sum_{r \in Q_1} \left[(a_{11}X_{1r1} - \ln(e^{a_{11}X_{1r1}} + e^{a_{12}X_{1r2}})\right] + \sum_{r \in Q_2} \left[a_{12}X_{1r2} - \ln(e^{a_{11}X_{1r1}} + e^{a_{12}X_{1r2}})\right] \qquad (1.155)$$

By substituting in the passenger's estimated values we get :

$$\bar{L} = \left[26a_{11} - \ln(e^{9a_{11}} + e^{7a_{12}}) - \ln(e^{8a_{11}} + e^{6a_{12}}) - \ln(e^{9a_{11}} + e^{8a_{12}})\right] + \left[36a_{12} - \ln(e^{7a_{11}} + e^{10a_{12}}) - \ln(e^{8a_{11}} + e^{9a_{12}}) - \ln(e^{5a_{11}} + e^{7a_{12}}) - \ln(e^{9a_{11}} + e^{10a_{12}})\right] \qquad (1.156)$$

By finding the partial derivatives $\frac{\partial \bar{L}}{\partial a_{11}}$ and $\frac{\partial \bar{L}}{\partial a_{12}}$ and setting them equal to zero, we get :

$$26-9\cdot\frac{e^{9a_{11}}}{e^{9a_{11}}+e^{7a_{12}}}-8\cdot\frac{e^{8a_{11}}}{e^{8a_{11}}+e^{6a_{12}}}-9\cdot\frac{e^{9a_{11}}}{e^{9a_{11}}+e^{8a_{12}}}-7\cdot\frac{e^{7a_{11}}}{e^{7a_{11}}+e^{10a_{12}}}-$$

$$-8\cdot\frac{e^{8a_{11}}}{e^{8a_{11}}+e^{9a_{12}}}-5\cdot\frac{e^{5a_{11}}}{e^{5a_{11}}+e^{7a_{12}}}-9\cdot\frac{e^{9a_{11}}}{e^{9a_{11}}+e^{10a_{12}}}=0 \qquad (1.157)$$

$$-7\cdot\frac{e^{7a_{12}}}{e^{9a_{11}}+e^{7a_{12}}}-6\cdot\frac{e^{6a_{12}}}{e^{8a_{11}}+e^{6a_{12}}}-8\cdot\frac{e^{8a_{12}}}{e^{9a_{11}}+e^{8a_{12}}}+36-10\cdot\frac{e^{10a_{12}}}{e^{7a_{11}}+e^{10a_{12}}}$$

$$-9\cdot\frac{e^{9a_{12}}}{e^{8a_{11}}+e^{9a_{12}}}-7\cdot\frac{e^{7a_{12}}}{e^{5a_{11}}+e^{7a_{12}}}-10\cdot\frac{e^{10a_{12}}}{e^{9a_{11}}+e^{10a_{12}}}=0 \qquad (1.158)$$

The system of equations (1.157) - (1.158) can only be solved by iterative numerical procedures. Newton's method (see the example in Scheid[139]) is used to solve these equations.

We get :

$$a_{11} \approx a_{12} \approx 10{,}4 \qquad (1.159)$$

The number of probabilities p_{rs} that are to be calculated in choice models based on disaggregated data is very large and equals the product of the number of polled passengers and the number of possible transportation modes. The other extreme in relation to models based on disaggregated data is that of choice models based on aggregated data where it is assumed that the variables in the choice function are equal for all passengers. This also means that the probability of choosing a certain mode of transportation is constant, and does not vary from passenger to passenger. In other words, for aggregated choice models :

$$p_{rs} = p_s \qquad r = 1,2,\ldots,n \tag{1.160}$$

where n is the number of passengers.

The likelihood function of a multinomial LOGIT model based on aggregated data reads :

$$L = \frac{n!}{n_1!n_2!\ldots n_z!} \prod_{s=1}^{z} (p_s)^{n_s} \tag{1.161}$$

In this case :

$$\bar{L} = \sum_{s=1}^{z} n_s \ln p_s \tag{1.162}$$

As for the disaggregated models, parameters a_{is} are defined from the condition $\frac{\partial \bar{L}}{\partial a_{is}} = 0$ for $\forall\, a_{is}$.

Let us illustrate the calibration of a binomial aggregated LOGIT model using a simple hypothetical example. We assume that passengers taking a private trip on a certain route choose between two possible modes of transportation solely on the basis of the price of the ticket. Data on the cost of transportation are give in Table 1.12.

TABLE 1.12 Cost of transportation by mode

Transportation mode A_i	*Travel costs* c_i
A_1	5
A_2	10

Let there be $n=1000$ passengers with $n_1=700$ choosing transportation mode A_1, and $n_2=300$ choosing mode A_2. We assume that the choice function takes this shape :

$$V(i) = a \cdot c_i \tag{1.163}$$

where :

a - parameter to be estimated

c_i - cost of transportation mode A_i .

The probability of mode A_1 or mode A_2 being chosen is equal

$$p_1 = \frac{e^{5a}}{e^{5a} + e^{10a}} , \quad p_2 = \frac{e^{10a}}{e^{5a} + e^{10a}} \tag{1.164}$$

Based on equation (1.162) we can write :

$$\bar{L} = n_1 \ln p(1) + n_2 \ln p(2) = n_1 \ln \frac{e^{V(1)}}{e^{V(1)} + e^{V(2)}} +$$

or :

$$n_2 \ln \frac{e^{V(2)}}{e^{V(1)} + e^{V(2)}} \tag{1.165}$$

$$\bar{L} = n_1 V(1) + n_2 V(2) - (n_1 + n_2) \ln \left[e^{V(1)} + e^{V(2)} \right] \tag{1.166}$$

In our example :

$$\bar{L} = 700 \cdot 5a + 300 \cdot 10a - 1000 \ln \left[e^{5a} + e^{10a} \right] \tag{1.167}$$

By finding the derivative of $\frac{d\bar{L}}{da}$ and setting it equal to zero we get this equation :

$$1500 \cdot e^{5a} - 3500 \cdot e^{10a} = 0$$

which gives when solved :

$$a = -\ 0{,}1694$$

The corresponding probabilities of other modes of transportation being chosen equal, respectively :

$$p_1 = \frac{e^{5\cdot(-0{,}1694)}}{e^{5\cdot(-0{,}1694)} + e^{10\cdot(-0{,}1694)}} = 0{,}7$$

$$p_2 = \frac{e^{10\cdot(-0{,}1694)}}{e^{5\cdot(-0{,}1694)} + e^{10\cdot(-0{,}1694)}} = 0{,}3$$

We assume that a new mode of transportation joins the route in question with cost $c_3=7$. We then calculate the market share between modes of transportation. The probabilities of different modes of transportation being used are :

$$p_1 = \frac{e^{5a}}{e^{5a}+e^{10a}+e^{7a}} = \frac{e^{5(-0{,}1694)}}{e^{5(-0{,}1694)}+e^{10(-0{,}1694)}+e^{7(-0{,}1694)}} = 0{,}47$$

$$p_2 = \frac{e^{10a}}{e^{5a}+e^{10a}+e^{7a}} = \frac{e^{10(-0{,}1694)}}{e^{5(-0{,}1694)}+e^{10(-0{,}1694)}+e^{7(-0{,}1694)}} = 0{,}20$$

$$p_3 = \frac{e^{7a}}{e^{5a}+e^{10a}+e^{7a}} = \frac{e^{7(-0{,}1694)}}{e^{5(-0{,}1694)}+e^{10(-0{,}1694)}+e^{7(-0{,}1694)}} = 0{,}33$$

As we can see, by introducing a third mode of transportation on the route in question, the share of the first mode decreases from 70% of all passengers transported to 47%, and the second mode goes from 30% to 20%.

LOGIT models can be successfully used to analyze the effects of introducing airplanes on new routes, how air fare affects the airplane's market share of all passengers, etc. Kanafani and Sadoulet[70] effectively applied choice models when studying the choice of transportation class and tariffs in long-distance traffic over the North Atlantic. On the one hand, choice models based on disaggregated data better describe the passenger's choice of transportation mode. On the other, a very large number of probabilities are calculated for these models. Models based on aggregated data do not entail any difficulty in calculations, but they certainly deviate more from reality due to the assumption that all passengers behave in the same manner when choosing a mode of transportation. Transition models between disggregated and aggregated include those where certain groups of polled passengers are aggregated. After compiling the poll results, groups of passengers who gave similar evaluations to the attributes of different alternatives can be treated as a group of passengers where everyone gives the same evaluation. In this case, the choice function variables equal the mean value of the variables obtained by polling a certain group.

The group of polled passengers is therefore divided into several smaller groups of passengers who all choose their mode of transportation in the same way.

1.10. Concluding Remarks on Air Transportation Demand

The previous presentations discussed the basic characteristics of transportation demand and the supply of transportation services. A classification was made of demand estimation models: depending on whether or not the model includes competitive modes of transportation, demand estimation models are divided into multimode models and models that are not a function of the characteristics of alternative modes of transportation. Demand models were also divided into macroscopic and microscopic models. It was shown that multimode models are primarily used to estimate transportation demand on shorter routes, while demand in long-distance traffic is essentially estimated independently of other modes of transportation. A detailed analysis was made of models where transportation demand is a function of time and models where demand is a function of socio-economic characteristics. The discussion on trip distribution included entropy models, proportional models, mean trip length models and gravity models. In the study of multimode models, special attention was given to abstract mode models, and the basic characteristics of the modal split were given, with a detailed analysis of the LOGIT model.

Transportation demand data are the main inputs when planning an air traffic network, planning needed transportation facilities, building new and reconstructing existing airports. Estimating transportation demand should be given considerable attention and the models presented herein provide the starting point for satisfactory estimates of air transportation demand.

2. MATCHING TRANSPORTATION FACILITIES AND PASSENGER DEMAND IN AIR TRAFFIC

2.1. Matching Transportation Facilities and Passenger Demand in Air Traffic

The main problem facing air carriers is that of coordinating transportation facilities and passenger demand. An entire range of highly disparate factors are involved in this coordination which is much more complex on an entire airline network than on individual routes. The output of coordinating transportation facilities and passenger demand is a daily and weekly flight schedule on different routes and a corresponding departure schedule during the day. Flight frequency on individual routes and airplane departure schedules are a function of both the size and distribution of passenger demand during the day, week and month, and of the organizational-technical facilities of the air carrier. The air carrier's organizational-technical facilities include the number of different types of aircraft in the fleet, the number and professional level of the flight, technical and commercial personnel, technical base facilities, the reservation system, etc. An entire range of problems arises when coordinating transportation facilities and passenger demand comprised, in addition to determining flight frequency and departure times, of rotating aircraft in the network, planning fleet maintenance, planning crew schedules, etc.

2.2. The Importance of Flight Frequency Determination

With the development of air traffic, methods and models to determine flight frequency and departure times also developed. However, constant improvement must be made to existing models and new models must be created to solve this problem. It has already been underscored that determining flight frequency is a very important step towards the completion of other planning activities that the air carrier must accomplish if he wishes to fly on an air network the best way possible. Flight frequency and airplane departure times are one of the most important factors involved in the passenger's decision to use an air carrier's services when there are several carriers flying on the same route. Johnston et al [69] studied the factors that determine the passenger's choice of air carrier when there is competition. In order to establish the importance of different criteria when choosing an air carrier, they polled 150 passengers on routes in Canada. Passengers evaluated the following criteria : flight schedule, safety, air fare, the characteristics of the airplane, the reservation system and service in the airplane and at the airport. The airline schedule criteria included these components : take off and landing on time, flight frequency and the existence of non-stop flights. Survey results indicated that business passengers consider the most important criterium to be the airline schedule, including flight frequency, that make them opt for a certain carrier. For passengers on private trips this criterium was in third place. The effect of flight frequency on the percentage share of a carrier out of the total number of passengers on routes with competition were also studied by Renard[126], Gelerman and de Neufville[48] and Ghobrial[54]. These papers, which will be discussed later on, also showed the extremely large influence of flight frequency on the air carriers' market share.

Flight frequency and aircraft departure times are also extremely important factors that influence the potential passenger modal split. This is particularly true for shorter and medium distance routes where there is very strong competition from all modes of transportation (automobile, bus, train, airplane).

A survey of models to estimate the number of passengers in air traffic was given in the previous chapter.

As mentioned in the previous chapter, the most important factors that determine a passenger's choice of air carrier are travel outlays and the travel time of different modes of transportation, flight frequency and aircraft departure times, bus and train departure schedules, respecting departure schedules, etc. The number of passengers that take the airplane on a route greatly depends on flight frequency and departure times. Proper insight into passengers' transportation demand and the distribution of this demand over time can increase the number of air transportation passengers by offering an adequate number of flights at suitable times. On the other hand, if the market offers a frequency that is too small or unsuitable airplane departure times during the day, it is clear that potential air transportation users will go over to other modes of transportation. There are some air carriers that do not pay enough attention to the effects that flight frequency and departure times have on the number of passengers they transport. Simpson[146] notes that the effects of flight frequency and schedules on transportation demand are very rarely examined. An investigation into the effects of flight frequency and the distribution of departure times on transportation demand must be given the attention it is due, particularly on routes where there is competition from several different modes of transportation.

2.3. The Concept of Passenger Flows and Changes in Their Intensity Over Time

Passenger flow can be defined, as mentioned before, as the number of passengers in a unit of time traveling from one city to another. We can write it as :

$$h_{ij}(t)=\frac{\begin{bmatrix}\text{the number of passengers traveling from city } i\\ \text{to city } j \text{ during time interval } (t,\ t+dt)\end{bmatrix}}{dt} \quad (2.1)$$

or :

$$h_{ij}(t)=\frac{d\begin{bmatrix}\text{the total number of passengers traveling from}\\ \text{city } i \text{ to city } j \text{ during time interval}(o,t)\end{bmatrix}}{dt} \quad (2.2)$$

where $h_{ij}(t)$ is the passenger flow between city i and city j at time t .

We also underscored earlier that potential passenger flow comprises the number of passengers in a unit of time that wish to travel from one city to another. For one reason or another a difference is very often found between the actual and potential passenger flow. For example, due to low frequency in flights between two cities, an air carrier operating without competition is not able to satisfy the market demand which means that the actual passenger flow is below its potential.

Passenger flow is a value that changes over time. Changes are noticed by month, by week, by day in the week, and finally by hour in the day. Monthly changes are of interest for their global view of the scope of traffic. In terms of monthly changes in passenger flows, most air routes can be divided into "business" or "tourist" routes. Business routes are characterized by relatively even monthly flows during the year, while tourist routes usually have seasonal "peaks" during the summer months. It is clear that daily or weekly flight frequencies on a route are a function of passenger flows, which means that monthly flow changes must be monitored and corresponding changes made in flight frequencies.

It is extremely important to monitor passenger flows by day in the week and particularly by hour in the day in order to solve the problem of determining flight frequency and departure times.

Fluctuations in passenger flows by day in the week were studied by Swan[15] and Powell[12]. By studying the fluctuation in transportation demand on a route with two competitive air carriers, Powell assumed that the total daily number of passengers wanting to travel V_o is a random variable subject to a normal probability distribution with mean μ and variation σ^2. Demand variability is conditioned above all by the fact that each passenger chooses by himself, independent of all others, both the day of his trip and the specific time interval during that chosen day. In addition to this, an entire series of other factors affect the variability of transportation demand, making a normal probability distribution logical. A normal distribution appears even before treating the number of passengers wishing to travel on a certain flight as a random variable. Since it is not possible (or quite difficult) to measure the number of passengers wanting to take a certain flight, the distribution of the number of passengers in the airplane is often used in place of this number. Swan[151] and Soumis[149] showed that this is also a normal probability distribution.

The papers of Gagnon[42], Hyman and Gordon[67], Simpson[146], Miller[104], Soumis et al[149] and Teodorovic[155], indicate the characteristic look of a passenger flow as a function of time during the day. Figure 2.1 shows the passenger flow's characteristic functional dependence on the time of day.

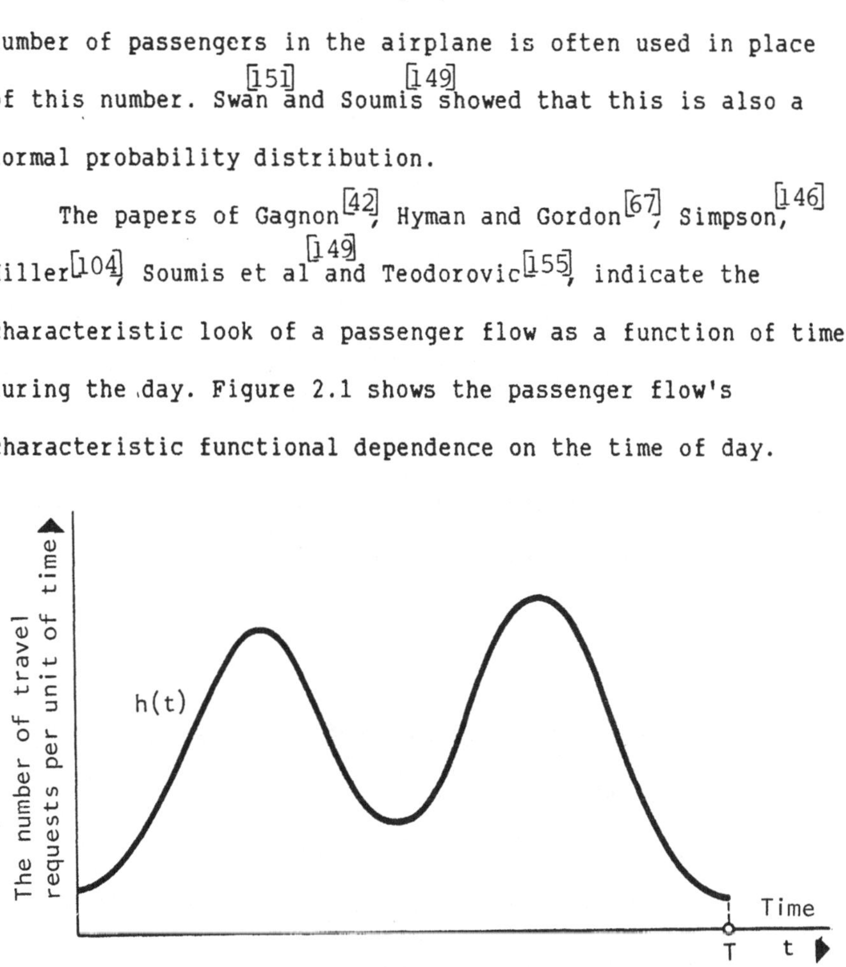

FIGURE 2.1. The number of travel requests per unit of time as a function of the time of day.

The passenger flow on Figure 2.1 is denoted by $h(t)$ and T indicates the time interval in which passengers expressed the desire to travel. The functional dependence between passenger flow and the time of day typically has two maximums or two peaks - in the morning and in the evening. The shape of the flow is obtained either by recording the arrival time of passengers when there is no fixed schedule, or by polling passengers when the schedule is known in advance.

2.4. The Passenger Load Factor

The average load factor is the ratio between the average number of passengers in the airplane and the number of seats it contains. Average load factor η_p during a fixed period of time on a specific route is defined as :

$$\eta_p = \frac{\lambda}{n \cdot N} \tag{2.3}$$

where :

λ - average number of passengers during a fixed period of time on the route,

n - number of seats in the airplane,

N - flight frequency, i.e. the number of flights during the time period under observation.

Swan [151] defines η_p as :

$$\eta_p = \frac{\mu}{n} \tag{2.4}$$

calling it the nominal load factor whereby μ is the mean of the number of passengers on the flight.

Equation (2.3) provides us with information on the significance of achieving the load factor which is one of the basic indicators of an air carrier's economic results. All air carriers try to achieve a load factor that is not less than the break even load factor in which revenues equal total outlays. For constant average passenger number λ on the route, the average passenger load factor η_p is a monotonic declining function of flight frequency N (Figure 2.2).

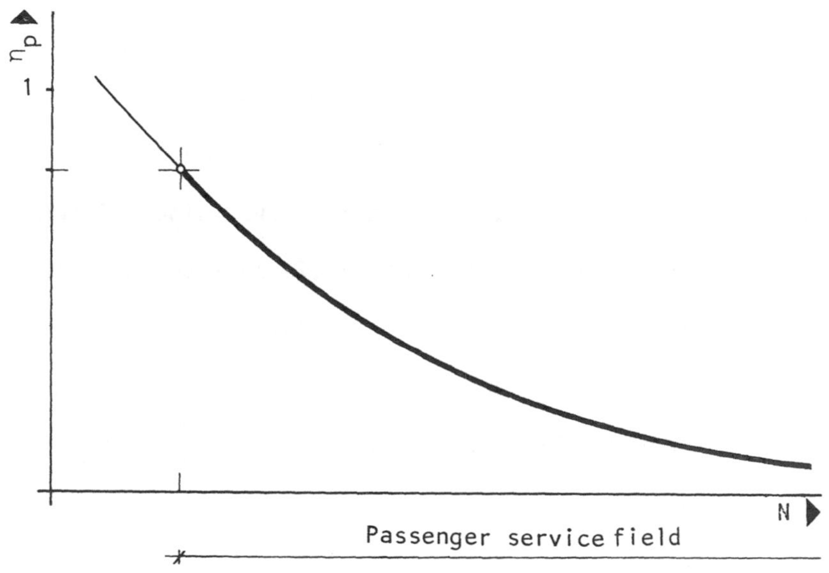

FIGURE 2.2. Passenger load factor as a function of flight frequency

The average load factor can serve as a rough indicator of the quality of the transportation services provided. To this effect, a rise in the average load factor increases the number of passengers that cannot find available seats on certain flights. It can be concluded that average load factor η_p is a very important indicator that is conditioned by flight frequency. This coefficient gives us orientational information about the air carrier's business results and the quality of its transportation services.

2.5. Calculating the number of refused passengers

It has already been mentioned that a route with a greater passenger load factor means that a certain number of passengers cannot find an available seat on the flight of their choice. Consideration of the behavior of these passengers logically points to some of their alternatives :

- refused passengers cancel their trip,
- refused passengers try to find a seat on the next flight,
- refused passengers try to find a seat on the first, second, third, ..., following flight,
- refused passengers try to find a seat on the closest flight (this means either on the next

flight or on the first preceding flight, by agreeing to travel earlier than desired).

Soumis et al[149] studied the behavior of passengers when choosing their flights and developed a model that distributes the entire passenger population by flight (MAPUM). They introduced a route attraction factor that is a function of take off time, the number of stopovers, the number of possible linkups to continue the trip and the quality of the cabin service. Results obtained by the MAPUM model were in good agreement with the actual number of passengers on different AIR CANADA flights.

The concept of "flight attraction" is described by Gagnon in the following way. Let d_1 and d_2 be the departure times of two consecutive flights, and a_1 and a_2 the arrival times at the chosen airport (Figure 2.3) where the attraction of the first flight ends and the attraction of the second flight begins at time t :

$$t = d_1 + \frac{a^2}{a+d} \tag{2.5}$$

where :

$$a = a_2 - a_1 \tag{2.6}$$

$$d = d_2 - d_1 \tag{2.7}$$

When both flights last the same time we have $a = d$, i.e. :

$$t = d_1 + \frac{a^2}{a+d} = d_1 + \frac{d^2}{2d} = d_1 + \frac{d}{2} =$$

$$= d_1 + \frac{d_2 - d_1}{2} = \frac{d_1 + d_2}{2} \tag{2.8}$$

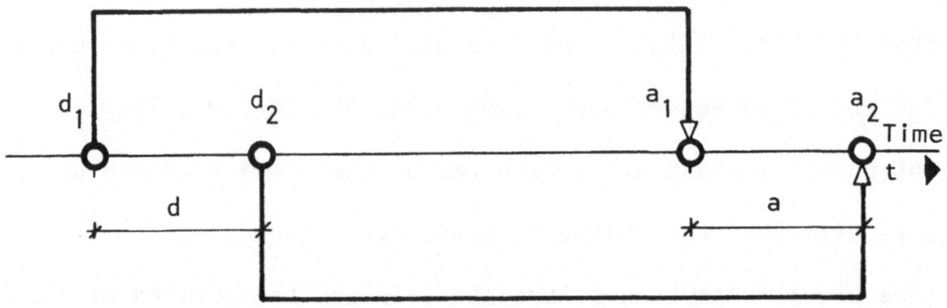

FIGURE 2.3. Take off and landing times of two consecutive flights

In this case, the passenger will adjust his desired travel time to the departure schedule so that he leaves as close as possible to the desired time. Teodorović [155] used the same assumption.

Every passenger makes his own decision independently of all other passengers concerning the day and time he wishes to travel. Should he not find an available seat on his desired flight, the passenger, as already mentioned, will cancel his trip, choose only the next flight, choose all other flights until he finds an available seat, etc. It is clear that numerous factors influence the total number of passengers on a flight. The probability theory and mathematical statistics tell us that when numerous independent random factors affect a phenomenon with each one having only an insignificant effect, then the phenomenon's characteristics are subject to a normal probability density. So it is logical to expect that the number of passengers on a flight is a random variable distributed by the normal probability density function. Soumis et al [149] cite that an analysis of the quarterly data on AIR CANADA routes confirmed this assumption. Swan [151] also demonstrated the normal probability density of the passenger load factor. The number of passengers on a flight indicates the transportation demand for that flight. Since it is rather difficult to measure transportation demand, knowing the passenger flight distribution provides the basic information to analyze demand and calculate the probability that a passenger cannot find a seat on his desired flight. As Swan indicates, the best measure of transportation service quality is the probability that a passenger cannot find a free seat on his desired flight. This probability is calculated in the following manner. We denote by X a random variable representing the number of passengers on the flight in question. Due to the previously introduced probability density function $f(x)$ is:

$$f(x) = \frac{1}{\sigma\sqrt{2\pi}} e^{-\frac{(x-\mu)^2}{2\sigma^2}} \qquad (2.9)$$

where μ and σ are the mean and the standard deviation of random variable X .

We denote by Y the random variable representing the number of passengers that could not find seats on their desired flights. The mean of this random variable is :

$$M(Y) = \int_n^\infty (X-n)\cdot f(x)dx \qquad (2.10)$$

where n is the capacity, i.e. the number of seats in the airplane.

With further transformations we get :

$$M(Y) = \int_n^\infty (X-\mu)f(x)dx + \int_n^\infty (\mu-n)f(x)dx \qquad (2.11)$$

and :

$$M(Y) = \int_n^\infty (x-\mu)f(x)dx + (\mu-n)\left[\Phi(\infty)-\Phi(\frac{n-\mu}{\sigma})\right] \qquad (2.12)$$

where

$$\Phi(t) = \frac{1}{\sqrt{2\pi}}\int_0^t e^{-\frac{u^2}{2}}du \qquad (2.13)$$

is the Laplace function whereby :

$$\Phi(\infty) = 0,5$$

After solving the integral in equation (2.12) we get :

$$\int_n^\infty (x-\mu)\cdot\frac{1}{\sigma\sqrt{2\pi}}e^{-\frac{(x-\mu)^2}{2\sigma^2}}dx = \frac{\sigma}{\sqrt{2\pi}}e^{-\frac{(n-\mu)^2}{2\sigma^2}} \qquad (2.14)$$

Finally, we can write the mean of the number of passengers that cannot find a place on their desired flight :

$$M(Y) = \frac{\sigma}{\sqrt{2\pi}} e^{-\frac{(n-\mu)^2}{2\sigma^2}} + (n-\mu)\left[\Phi\left(\frac{n-\mu}{\sigma}\right) - 0,5\right] \quad (2.15)$$

Table 2.1. calculates the mean of the number of passengers refused on an airplane with 115 seats and different parameters μ and σ .

The results obtained are completely logical and expected. For the same demand variability (for constant σ), an increase in the average number of passengers in the airplane shows an increase in the mean of the number of passengers that cannot find a place on their desired flight. Also, for a fixed average number of passengers in the plane (constant μ), a rise in the demand variability (a rise in σ) considerably increases the mean of the number of passengers refused. This observation is particularly important since it indicates that when analyzing the utilization level of transportation facilities, special attention must be given to demand variability or the deviation of the number of passengers on certain flights from the average number of passengers in the airplane, in addition to analyzing the passenger load factor. Based on equation (2.15) which establishes the average number of passengers per flight, further analyses can be made concerning changes in the flight frequencies on the route in question, changes in the type of aircraft (changing the capacity), etc. In this way, note is

TABLE 2.1. Mean of the number of passengers refused per flight for an airplane with 115 seats

No.of seats *n*	Mean No. of passengers μ	Standard deviation of the No. of passengers σ	Mean No. of refused passengers per flight *M(Y)*
115	70	10	<1
115	70	20	<1
115	70	30	<1
115	70	40	2,614
115	80	10	<1
115	80	20	<1
115	80	30	1,825
115	80	40	4,352
115	90	10	<1
115	90	20	<1
115	90	30	3,377
115	90	40	6,47
115	100	10	<1
115	100	20	2,625
115	100	30	5,937
115	100	40	9,56
115	110	10	1,977
115	110	20	5,723
115	110	30	9,618
115	110	40	13,58

made of the advantages and defects of each alternative, which differ from each other by flight frequency and/or type of aircraft, and facilitates a suitable decision concerning how to organize air traffic on a specific route.

2.6. Flight Frequency and Air Carrier Competition

When only one air carrier operates on a route, flight frequency, as we have seen, has a direct effect on both the economic results of transportation and on the quality of service. When an air carrier has a monopoly on the market, small and inadequate flight frequency or a high load factor can prompt an increase in the number of passengers refused and turn them to alternative modes of transportation. When there are several air carriers on the same route, it is even more important to accurately determine flight frequency and departure times in order to maximize the air carrier's market share of the total number of passengers.

Let us take a look at Figure 2.4, presenting the functional dependence between passenger flow and the time of day.

Let there be two air carriers. The first carrier has departures at times x_1, x_2, x_3, x_4 and x_5. The second air carrier has flights leaving at times x_1, x_2 and x_4. Let us consider the case when both carriers use aircraft with the same capacity and tariffs are the same for both on all flights. The attraction of individual flights is determined in this example with the assumption that the passenger will choose the flight with the time closest to his needs. Let the area of each of the 5 zones under curve $h(t)$ be equal. At

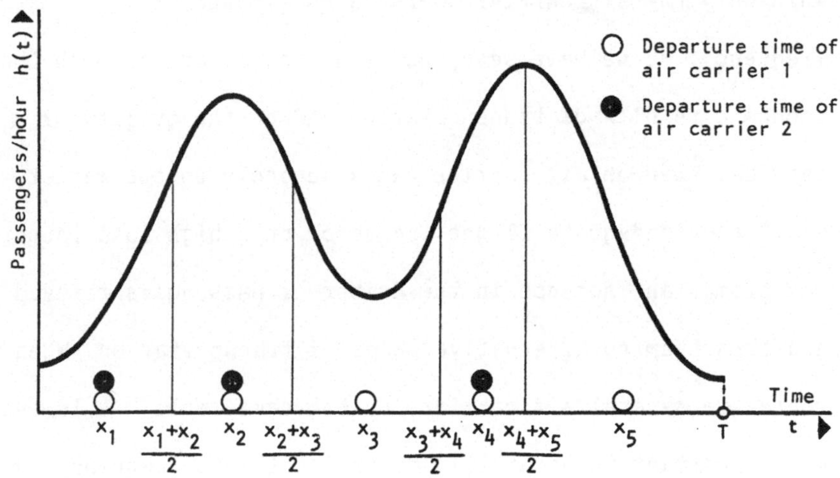

FIGURE 2.4. Departure times of competitive carriers

times x_3 and x_5 carrier 1 has no competition, so that he will transport all passengers wanting to travel within time intervals

$$\left(\frac{x_2+x_3}{2}, \frac{x_3+x_4}{2}\right) \text{ and } \left(\frac{x_4+x_5}{2}, T\right)$$

This means that with just these 2 flights air carrier 1 will transport $\frac{2}{5}$ of all passengers wishing to travel during time interval $(0,T)$. On flights departing at times x_1, x_2 and x_4 air carrier 1 will transport $\frac{1}{2} \cdot \frac{3}{5}$ of all passengers wishing to fly during time interval $(0,T)$. This means if air carrier 1 flies $\frac{5}{8}$ of all flights on this route, he will succeed in transporting $\frac{2}{5} + \frac{1}{2} \cdot \frac{3}{5} = \frac{7}{10}$ of all passengers. On the other hand, air carrier 2 flying $\frac{3}{8}$ of all flights will transport $\frac{1}{2} \cdot \frac{3}{5} = \frac{3}{10}$ of the total number of passengers. This simple example alone indicates that an air carrier with a greater flight frequency will succeed for the most part in gaining a percentage share of the number of passengers that is greater than his percentage share of the total number of flights. The explanation for this lies in the fact that with greater frequency, the air carrier does not have any competition at certain moments in time.

The problem of determining the percentage share of passengers as a function of the percentage share of total flights on a route was considered by numerous authors. Renard[126], Gelerman and de Neufville[48], de Neufville and King[24] and Powell[121] used the model shown by Renard that between the percentage share of the number of passengers MS_i and the percentage share of the number of flights FS_i there is a relationship having this shape :

$$MS_i = \frac{FS_i^{\alpha}}{\sum_{i=1}^{k} FS_i^{\alpha}} \qquad (2.16)$$

where :

MS_i - percentage share of air carrier i of the total number of passengers (market share),

FS_i - percentage share of air carrier i of the total number of flights (frequency share),

k - number of carriers operating on the route,

α - parameter that is set when calibrating the model.

As noted by Gelerman and de Neufville,[43] parameter α which is a function of the number of carriers operating on the route, is usually within the limits $1 < \alpha < 2$.

A typical market share of the number of passengers as a function of the frequency share is shown in Figure 2.5.

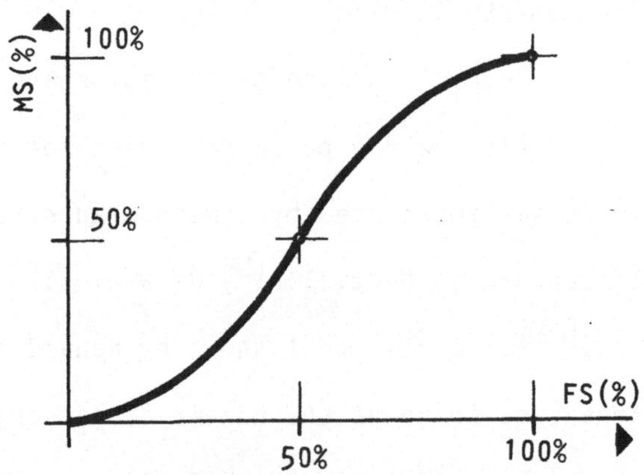

FIGURE 2.5. Market share of the passengers as a function of the frequency share of flights

Etschmaier[34] demonstrated that between the percentage share of passengers and the number of flights there can also be a relationship taking this shape:

$$MS_{js} = \frac{\sum_{i} \alpha_{ij} X_{ijs}}{\sum_{i} \alpha_{ijs} X_{ijs} + \sum_{i,r} \beta_{ijr} f_{ijr}} \qquad (2.17)$$

where :

MS_{js} - the market share of carrier s of all passengers transported on route j ,

X_{ijs} - carrier s frequency on route j with aircraft type i ,

f_{ijr} - competitive carrier r flight frequency on route j with aircraft type i ,

$\alpha_{ijs}, \beta_{ijr}$ - parameters estimated statistically.

Equation (2.17) refers therefore to the case when there are competitive carriers on the network whose fleets have aircraft with different capacities.

Equations (2.16) and (2.17) give the carrier an idea of the effects that could be achieved by changing flight frequency and/or the type of aircraft on the route.

2.7. Flight Frequency and the Quality of Service

A logical question that arises when discussing the problem of flight frequency determination is the effect flight frequency has on the quality of transportation services. This question was considered by Gordon and de Neufville[58], Douglas and Miller[30], Eriksen[32], Swan[151], Kanafani and Ghobria[173], Ghobria[154] and Teodorovic[155]. Time losses caused by the airline schedule usually comprise the absolute value of the time difference between the desired departure time and the time that is actually possible considering the airline schedule. For example, if a passenger wants to travel at 1:00 p.m. and the flight schedule will only let him travel at 5:00 p.m., this time difference is 4 hours. It is clear that a rise in flight frequency and an increase in the number of possible departures offered to travellers will decrease average time losses per passenger caused by the flight schedule. The literature gives numerous formulas to calculate schedule delay. These will be compared later on when discussing the problem of determining flight frequency on a route.

We noted earlier that due to the variability in air traffic demand the number of passengers on a flight is a random variable that is most often distributed by a normal probability density function. This results in a number of passengers that cannot find a seat on the flight of their choice. It was shown that this number increases with a rise in the load factor. This means that low flight frequency suitable to a high load factor increases the number of refused passengers that must choose another flight. These passengers must therefore deviate even more from their desired departure time since there is no room on the flight that is the closest to this time. The difference between the departure time of the flight on which the passenger can find a seat and the departure time of the flight that the passenger first chose is called stochastic delay. In order to calculate the average stochastic delay per passenger, the probability must first be calculated that the passenger cannot find a vacant seat on the flight of his choice.

A similar time difference appears when the passenger has a proper reservation for the flight of his choice, but is unable to travel due to overbooking. The problem of overbooking was studied by, among others, Falkson[39], Rothstein[132-134], Shlifer and Vardi[145], Simonl[141-143], Vickrey[167] and Nagarajan[107]. Nagarajan showed with the use of passenger surveys that passengers do not demand a large sum of money in compensation for the stochastic time loss caused by overbooking that makes them wait for some later flight. The figures Nagarajan arrived at when studying the problem of overbooking are probably similar to the waiting times caused by a high load factor, i.e. low flight frequency.

As we can seen, flight frequency is one of the factors that most influences the quality of transportation services since it is directly responsible for both time losses caused by the flight schedule and stochastic time losses. The size of these losses provides us with much more information about the quality of transportation services than the flight frequency figure.

2.8. Determining Flight Frequency on a Route That Maximizes Air Carrier Profit

In addition to flight frequency's effect on the quality of transportation services, it also has an essential impact on the number of passengers, total revenues and total direct and indirect operating costs. Using these facts as a point of departure De Vany and Garges [29] developed a model to determine flight frequency that maximizes air carrier profit for a route with no competition. They expressed total profit P on a route between city i and city j in the form :

$$P = c \cdot \lambda_{ij}(c,N) - DOC \cdot N - L \cdot N - IOC \cdot \lambda_{ij}(c,N) - IC \qquad (2.18)$$

where :

c - the air fare from city i to city j ,

N - the number of flights between city i and city j during a fixed period of time (flight frequency),

$\lambda_{ij}(c,N)$ - the number of passengers between city i and city j for air fare c and flight frequency N,

DOC - direct operating costs per flight ,

L - the landing tax,

IOC - indirect operating costs (without landing tax) per passenger,

IC - indirect costs.

The flight frequency that maximizes profit is determined from the condition :

$$\frac{\partial P}{\partial N} = 0 \qquad (2.19)$$

whence this equation :

$$\frac{\partial P}{\partial N} = c\,\frac{\partial \lambda}{\partial N} - DOC - L - IOC\,\frac{\partial \lambda}{\partial N} = 0$$

After recomposition, we get :

$$N = \frac{C-IOC}{DOC+L} \cdot \lambda \cdot \frac{\frac{\partial \lambda}{\lambda}}{\frac{\partial N}{N}} \qquad (2.20)$$

Expression $e = \frac{\frac{\partial \lambda}{\lambda}}{\frac{\partial N}{N}}$ is the elasticity of the number of passengers compared to changes in frequency, so we can write:

$$N = \frac{(c-IOC)\cdot\lambda\cdot e}{DOC+L} \qquad (2.21)$$

The average load factor of airplane η_p that maximizes profit is :

$$\eta_p = \frac{\lambda}{n\cdot N} = \frac{DOC+L}{(c-IOC)\cdot e\cdot n} \qquad (2.22)$$

In order to practically use equations (2.21) and (2.22), value e on the route must be known. Should there be suitable statistical data on changes in flight frequency and corresponding changes in the number of passengers, using equations (2.21) and (2.22) will give air carriers complete information on the flight frequency that maximizes profit on the route.

Swan also considered the problem of determining flight frequency on a route that maximizes air carrier profits under conditions with no competition. When calculating the daily frequency that maximizes profits, Swan[151] expressed total daily costs for flights between two cities as :

$$C = c_1 \cdot S + c_2 \cdot N + c_3 \cdot \lambda \qquad (2.23)$$

where :

N - daily flight frequency on the route,

C - total daily costs of operating air transportation on the route,

S - total number of seats available during the day,

λ - average daily number of passengers on the route,

c_1- costs per available seat,

c_2- costs per take off,

c_3- costs per passenger.

Total revenues R realized by the carrier are :

$$R = c \cdot \lambda \qquad (2.24)$$

Total profit P is simply obtained as the difference between total revenues and total costs :

$$P = R - C \qquad (2.25)$$

or :

$$P = c \cdot \lambda - c_1 \cdot S - c_2 \cdot N - c_3 \cdot \lambda \qquad (2.26)$$

Since the average load factor is η_p :

$$\eta_p = \frac{\lambda}{S} \tag{2.27}$$

this is

$$P = c \cdot \lambda - c_1 \cdot \frac{\lambda}{\eta_p} - c_2 \cdot N - c_3 \cdot \lambda \tag{2.28}$$

In his work, Swan used the assumption that transportation demand on the route can be shown by this equation :

$$\lambda = k_1 \cdot PP^{\alpha} \tag{2.29}$$

where :

PP - perceived price for making the trip,

α, k_1 - parameters that are calibrated.

With the assumption that PP can be presented in the form :

$$PP = c+v\left[t_B + f_1(N) + f_2(\eta_p, N)\right] \tag{2.30}$$

where :

v - monetary value for the time passengers lose,

t_B - block time on the route (time from motor start-up at the origin airport to motor shut-down at the destination airport),

$f_1(N)$ - schedule delay per passenger,

$f_2(\eta_p, N)$ - stochastic delay per passenger,

we get :

$$\lambda = k_1 \left\{c+v\left[t_B+f_1(N)+f_2(\eta_p, N)\right]\right\}^{\alpha} \tag{2.31}$$

or :

$$P = k_1\left\{c+v\left[t_B + f_1(N) + f_2(\eta_p, N)\right]\right\}^{\alpha} \cdot \left[c - \frac{c_1}{\eta_p} - c_3\right] - c_2 \cdot N \tag{2.32}$$

From condition $\frac{dP}{d\bar{c}} = 0$ we get the price of air fare that maximizes the air carrier's profit :

$$c = \frac{c_1}{\eta_p} + c_3 - \frac{1}{\alpha} \cdot PP \qquad (2.33)$$

This means that equation (2.33) defines the price of the ticket that maximizes the air carrier's profits. It can be further transformed as follows :

$$c = \frac{c_1}{\eta_p} + c_3 - \frac{1}{\alpha}\{c+v\left[t_B+f_1(N)+f_2(\eta_p,N)\right]\} \qquad (2.34)$$

and :

$$c = \frac{\alpha}{\alpha+1}\{\frac{c_1}{\eta_p} + c_3 - \frac{1}{\alpha}\cdot v\left[t_B+f_1(N)+f_2(\eta_p,N)\right]\} \qquad (2.35)$$

The frequency that maximizes profit is obtained from this condition :

$$\frac{dP}{dN} = 0 \qquad (2.36).$$

After derivation we get :

$$\frac{d\lambda}{dN}(c - \frac{c_1}{\eta_p} - c_3) - c_2 = 0 \qquad (2.37)$$

Since $\lambda = k_1 PP^{\alpha}$, this is :

$$\frac{d\lambda}{dN} = \frac{\alpha}{PP} \cdot \lambda \cdot \frac{dPP}{dN} \qquad (2.38)$$

By substituting from (2.38) into (2.37) we get :

$$\frac{\alpha \cdot \lambda}{PP \cdot c_2}\left[c - \frac{c_1}{\eta_p} - c_3\right] \cdot \frac{dPP}{dN} = 1 \qquad (2.39)$$

Since : $PP = c+v\left[t_B+f_1(N)+f_2(\eta_p,N)\right]$

then :

$$\frac{dPP}{dN} = v\{\frac{d[f_1(N)]}{dN} + \frac{d[f_2(\eta_p, N)]}{dN}\} \qquad (2.40)$$

Expression $\frac{dPP}{dN}$ figures in equation (2.39) that should calculate the flight frequency that maximizes the air carrier's profit. On the other hand, we see (based on equation (2.40)) that this expression depends on

$\frac{d[f_1(N)]}{dN}$ and $\frac{d[f_2(\eta_p, N)]}{dN}$

or on the way that time losses are defined.

Swan calculated schedule delay using this equation:

$$f_1(N) = \frac{5,7}{N} \qquad (2.41)$$

It is assumed that transportation demand is expressed 22.8 hours a day and that the average time loss caused by the flight schedule equals $\frac{1}{4}$ of the time interval between two take offs, which is $\frac{22,8}{N}$.

Swan also showed that stochastic delay per passenger can be approximated as follows :

$$f_2(\eta_p, N) = \frac{57 \cdot (\frac{\mu}{n})^9}{N} \qquad (2.42)$$

where μ is the average number of passengers on the flight and n the airplane capacity.

If, when calculating average stochastic delay, the average time interval between two flight is multiplied by the probability that the passenger cannot find a seat on the first flight of his choice, we get :

$$f_2(\eta_p, N) = p_1 \cdot \frac{22,8}{N} \qquad (2.43)$$

or :

$$f_2(\eta_p, N) = \frac{22,8}{N} \left[0,5 - \Phi\left(\frac{n-\mu}{\sigma}\right) \right] \qquad (2.44)$$

By finding the corresponding derivatives $\frac{d\left[f_1(N)\right]}{dN}$ and $\frac{d\left[f_2(\eta_p, N)\right]}{dN}$ and substituting them in equation (2.40) we get :

$$\frac{dPP}{dN} = v \left[- \frac{5,7}{N^2} - \frac{57 \cdot \left(\frac{\mu}{n}\right)^9}{N^2} \right] \qquad (2.45)$$

By substituting from (2.45) into (2.39) we get :

$$N = \sqrt{\frac{\lambda}{PP \cdot c_2}\left(c - \frac{c_1}{\eta_p} - c_3\right) \left\{ -\alpha v \left[5,7 + 57\left(\frac{\mu}{n}\right)^9 \right] \right\}} \qquad (2.46)$$

The use of equations (2.35) and (2.46) comprises a previous calibration of the demand model for air transportation services (determining parameters k_1 and α), an estimate of the value of time v , an estimate of parameters μ and σ characterizing the variability of transportation demand, and an estimate of costs c_1, c_2 and c_3 .

When discussing flight frequency on a route without competition whereby the air carrier maximizes his profit,

Swan also considered flight frequency in order to transport the greatest number of passengers. This frequency is determined with the condition that the price of transportation equals the average cost per passenger. This means that :

$$c = \frac{C}{\lambda} = c_1 \cdot \frac{S}{\lambda} + c_2 \cdot \frac{N}{\lambda} + c_3 \qquad (2.47)$$

or :

$$c = \frac{c_1}{\eta_p} + c_3 + c_2 \cdot \frac{N}{\lambda} \qquad (2.48)$$

On the other hand, based on equation (2.31) we have :

$$\lambda = k_1\{c+v\left[t_B+f_1(N) + f_2(\eta_p, N)\right]\}^{\alpha} \qquad (2.49)$$

By substituting from (2.48), (2.41) and (2.42) into (2.49) we get :

$$\lambda = k_1\left\{\frac{c_1}{\eta_p} + c_3 + c_2 \cdot \frac{N}{\lambda} + v\left[t_B + \frac{5,7}{N} + \frac{57 \cdot (\frac{\mu}{n})^9}{N}\right]\right\} \qquad (2.50)$$

From condition $\frac{d\lambda}{dN} = 0$ we get : (2.51)

$$N = \sqrt{\lambda} \cdot \sqrt{\frac{v\left[5,7 + 57(\frac{\mu}{n})^9\right]}{c_2}} \qquad (2.52)$$

Equation (2.52) therefore determines flight frequency on a route in order to transport the greatest number of passengers when there is no competition, with the price of transportation equal to the average cost per passenger.

2.9. An Analysis of the Effects of Flight Frequency on Average Airline Schedule Delay

Numerous expressions are used in professional literature to determine flight frequency on one route or all branches of a transportation network in order to calculate average time losses caused by airline schedules. As we have already seen, Swan,[151] when determining flight frequency on a route that maximizes the air carrier's profit, assumed that average schedule delay per passenger equals $\frac{1}{4}$ of the average time interval between two take offs. Ghobrial[54] considered the average schedule delay to be $\frac{4,5}{N}$. In some cases in the literature, average schedule delay is taken to be one-half the average time interval between two take offs. It is clear that the results that indicate the quality of transportation services, and thereby flight frequency on the route, vitally depend on how schedule delays are defined. Teodorovic[155] made a detailed analysis of the effects of flight frequency on the average airline schedule delay, using the following premises.

The time intervals between passengers' announcements of their wish to travel are random variables characterized by the probability density function. Figure 2.6 denotes by t_i ($i = 1,2,3,...$) the moments individual passengers indicate a desire to travel and by T_i ($i = 1,2,3,...$) the time intervals between these moments.

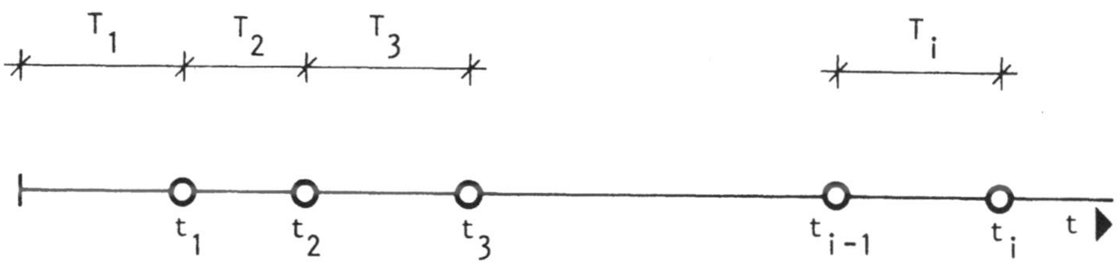

FIGURE 2.6 Time intervals between passengers' requests to travel

Random variables $T_1, T_2, T_3, \ldots,$ are independent of each other and are distributed according to the same probability density function. They are independent of each other because the reason one passenger wants to travel at a certain time is not linked to analogous reasons of other passengers. We denote by $f(t)$ the probability density function of random variables T_i that represent time intervals between individual passenger's requests to travel. The mean and standard deviation of these random variables are denoted by μ and σ, respectively. When determining average airline schedule delay, Teodorovic [155] assumed that the passenger would choose the departure time that was closest to his desired departure time. It is clear that average schedule delay depends both on flight frequency and on the departure times chosen during the day. At this point we will determine average airline schedule delay exclusively as a function of flight frequency, while the effects on this average delay of chosen departure times during the day will be discussed later on.

Figure 2.7 denotes by $x_1, x_2, \ldots, x_N$ airplane departures during time interval $(0, T)$.

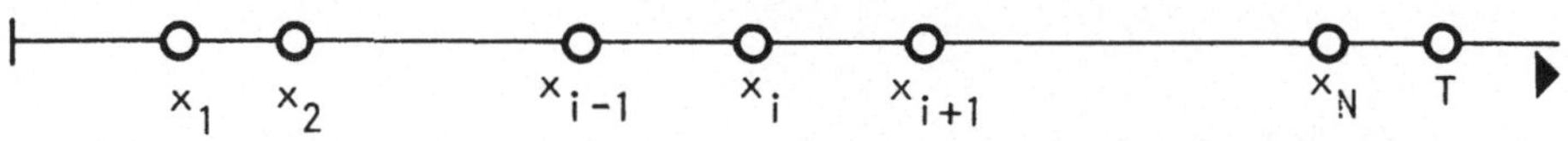

FIGURE 2.7 Airplane departures during time interval $(0, T)$

Let us note departure x_i. According to our earlier expressed premise that the passenger will choose the flight closest in time, all passengers will choose departure x_i who wish to travel during time interval

$$\left(\frac{x_{i-1} + x_i}{2}, \frac{x_i + x_{i+1}}{2}\right)$$

Let the average number of passengers on every flight on the route under observation be $2m$ and let m passengers wish to travel during interval $\left(\frac{x_{i-1} + x_i}{2}, x_i\right)$ and m passengers during interval $\left(x_i, \frac{x_i + x_{i+1}}{2}\right)$ The times that the $2m$ passengers wanted to travel are shown on Figure 2.8.

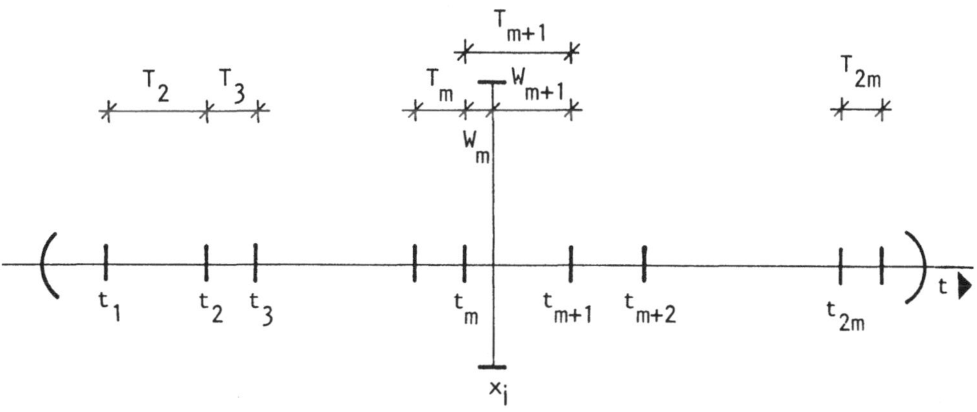

FIGURE 2.8 Passenger desired travel time

We denote by W_i the absolute time deviation between the actual and desired departure time of the i-th passenger. The figures for $W_i (i = 1,2,3,\ldots 2m)$ for individual passengers equals :

$$
\begin{aligned}
W_1 &= T_2 + T_3 + T_4 + \ldots + T_m + W_m \\
W_2 &= T_3 + T_4 + \ldots + T_m + W_m \\
W_3 &= T_4 + T_5 + \ldots + T_m + W_m \\
&\ldots \\
&\ldots \\
W_m & \\
W_{m+1} & \\
W_{m+2} &= T_{m+2} + W_{m+1} \\
W_{m+3} &= T_{m+3} + T_{m+2} + W_{m+1}
\end{aligned}
\qquad (2.53)
$$

....

....

$$W_{2m} = T_{2m} + T_{2m-1} + T_{2m-2} + \ldots + T_{m+2} + W_{m+1}$$

where : $W_m + W_{m+1} = T_{m+1}$

It is easy to show that random variable W that equals $W = \sum_{i=1}^{2m} W_i$ and represents the absolute total time deviation between the actual and desired departure times of all passengers who wanted to travel during time interval $(\frac{x_{i-1} + x_i}{2}, \frac{x_i + x_{i+1}}{2})$, equals the sum of m^2 random variables T_i that are mutually independent and have the same probability density function $f(t)$ with mean μ and standard deviation σ. Based on results from the probability theory we can state that random variable W is distributed according to the normal probability density function with mean $m^2\mu$ and standard deviation σm. Thus we have concluded that the probability density of the total schedule delay of all passengers who wanted to travel during time interval $(\frac{x_{i-1} + x_i}{2}, \frac{x_i + x_{i+1}}{2})$ equals :

$$g(w) = \frac{1}{\sigma m \cdot \sqrt{2\pi}} e^{-\frac{[w-m^2\mu]^2}{2\sigma^2 m^2}} \qquad (2.54)$$

The mean of random variable W is :

$$M(W) = m^2 \cdot \mu \qquad (2.55)$$

Average schedule delay per passenger s_d is :

$$s_d = \frac{M(W)}{2m} = \frac{m}{2} \cdot \mu \qquad (2.56)$$

The average number of passengers per flight is :

$$2m = n \cdot \eta_p \qquad (2.57)$$

It can be easily shown that :

$$s_d = \frac{n \cdot \eta_p}{4} \cdot \mu \qquad (2.58)$$

Passenger travel demand was expressed during time interval $(0, T)$. It is clear that :

$$\eta_p = \frac{\lambda}{n \cdot N} \qquad (2.59)$$

$$\mu = \frac{T}{\lambda} \qquad (2.60)$$

where N is flight frequency, n the number of seats in the airplane and λ the average number of passengers that wanted to fly during time interval $(0, T)$.

Based on equations (2.58), (2.59) and (2.60), we get :

$$s_d = \frac{T}{4N} \qquad (2.61)$$

where s_d is obtained in the same units as T .

Equation (2.61) provides very quick information on the quality of transportation services, as it is very easy to calculate the average delay between actual and desired departure times for a chosen flight frequency and known time interval during which passengers express the desire to travel.

In order for the air carrier to operate at a profit, the average load factor must be within these limits :

$$\frac{K}{n \cdot c} \leqslant \eta_p \leqslant 1 \tag{2.62}$$

where n is the number of seats in the airplane, c the air fare per passenger and K the total costs per flight (direct operating, indirect operating and indirect costs).

When the average passenger load factor satisfies equation (2.62), flight frequency N is within these limits :

$$\frac{\lambda}{n} \leqslant N \leqslant \frac{\lambda \cdot c}{K} \tag{2.63}$$

Based on equations (2.61) and (2.62) we can conclude that when an air carrier operates at a profit, the average time deviation per passenger between actual and desired departure times (average schedule delay) is within these limits :

$$\frac{n}{4\lambda} \cdot T \leqslant s_d \leqslant \frac{K}{4\lambda \cdot c} \cdot T \tag{2.64}$$

Relations (2.62), (2.63) and (2.64) create a link between flight frequency N, average load factor η_p that serves as a measure of the economic effects realized by the air carrier and average schedule delay per passenger s_d which serves as a measure of transportation service quality.

Average time losses are determined exclusively as a function of flight frequency. It is clear that these average time losses are a function of both flight frequency and of the chosen airplane departure times during the day. If the departure times are better adapted to the shape of the passenger flow, i.e. to function $h(t)$, average time losses will be smaller, and vice versa. It is evident that the same flight frequency can give different schedule delays, depending on the departure times chosen. We must therefore estimate the effect that departure times have on time losses. The size of this effect must first be discovered before equation (2.61) can be verified and conclusions made as to its usability.

Teodorovic [155] therefore inquired into the least possible schedule delay for a chosen flight frequency and compared it to the figure given by relation (2.61). This was done in the following manner.

The number of passengers $H(t)$ expressing a desire to travel up to some time $t \in (0,T)$ equals :

$$H(t) = \int_0^t h(t)dt \qquad (2.65)$$

Figure 2.9 presents function $H(t)$ and airplane departure times during the day $x_1, x_2, \ldots, x_N$.

According to the earlier adopted assumption that passengers choose the flight nearest in time, we have the

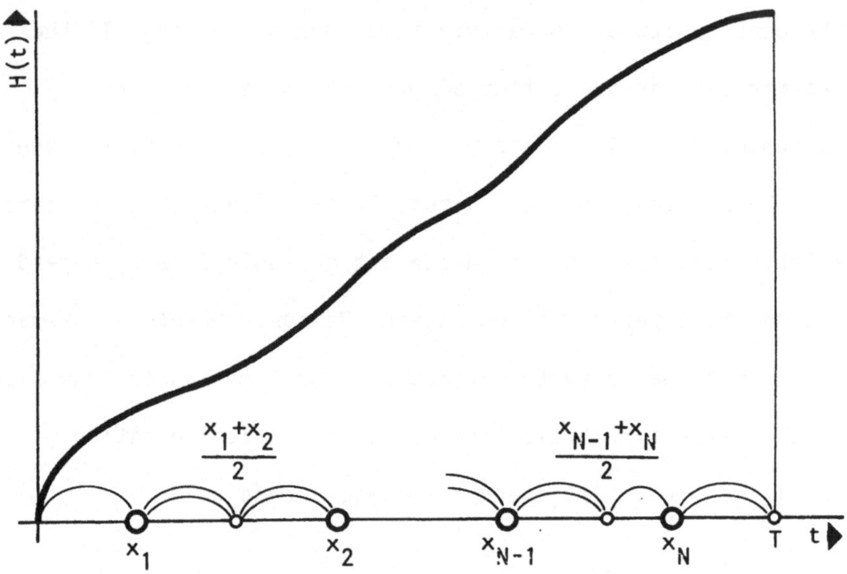

FIGURE 2.9 Passengers choose the flight nearest in time

following situation. The flight at time x_1 will be chosen by all passengers who wanted to travel during time interval $(0, \frac{x_1 + x_2}{2})$. The flight at time x_2 will be chosen by all passengers wanting to travel during time interval $(\frac{x_1 + x_2}{2}, \frac{x_2 + x_3}{2})$, etc. (Figure 2.10).

Total time losses for all passengers wanting to travel during time interval $(0, T)$ are shown in the shaded areas of Figure 2.10. Based on this, total time losses W^* equal :

$$W^* = 2\int_0^{x_1} H(t)dt + 2\int_0^{x_2} H(t)dt + \ldots +$$

$$+ 2\int_0^{x_N} H(t)dt - 2\int_0^{\frac{x_1+x_2}{2}} H(t)dt - 2\int_0^{\frac{x_2+x_3}{2}} H(t)dt - \ldots -$$

$$- 2\int_0^{\frac{x_{N-1}+x_N}{2}} H(t)dt - x_N \cdot H(T) -$$

$$- \int_0^{T} H(t)dt + T \cdot H(T) \qquad (2.66)$$

where :

$$H(T) = \lambda \qquad (2.67)$$

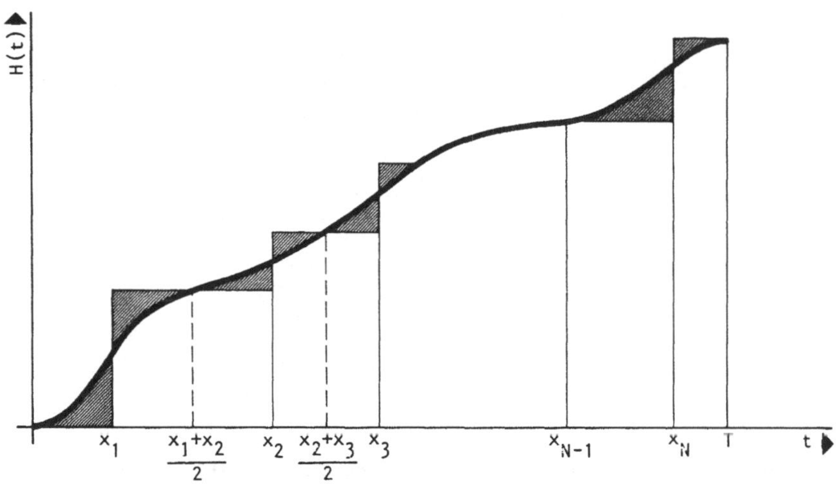

FIGURE 2.10 Total time losses for all passengers

Airplane departure times during the day $x_1, x_2, \ldots, x_N$ where total time losses W^* are minimal are obtained from the condition :

$$\frac{\partial W^*}{\partial x_i} = 0 \tag{2.68}$$

where $i = 1,2,\ldots,N$.

By finding the partial derivative we get that departure times $x_1, x_2, \ldots, x_N$ where the time difference is minimal must satisfy the system of equations :

$$H(x_1) = \frac{H\left(\frac{x_1 + x_2}{2}\right)}{2}$$

$$H(x_2) = \frac{H\left(\frac{x_1 + x_2}{2}\right) + H\left(\frac{x_2 + x_3}{2}\right)}{2}$$

$$H(x_3) = \frac{H\left(\frac{x_2 + x_3}{2}\right) + H\left(\frac{x_3 + x_4}{2}\right)}{2}$$

$$\ldots \quad \ldots\ldots\ldots \quad \ldots\ldots\ldots$$

$$\ldots\ldots \quad \ldots\ldots\ldots \quad \ldots\ldots\ldots \tag{2.69}$$

$$H(X_{N-1}) = \frac{H\left(\frac{X_{N-2} + X_{N-1}}{2}\right) + H\left(\frac{X_{N-1} + X_N}{2}\right)}{2}$$

$$H(X_N) = \frac{H\left(\frac{X_{N-1} + X_N}{2}\right) + H(T)}{2}$$

Based on equation (2.66) and (2.69) we can calculate the minimal total, i.e. average time loss per passenger for the already known shape of function $H(t)$. It is also possible using equation (2.66) to calculate total time losses for different values of $x_1, x_2, \ldots, x_N$. This lets us quantify the effects of the time schedule on the total or average time losses which will be used to verify the validity of equation (2.61).

In order to verify these results, the nature of the passenger flow must be previously studied, i.e. the shape of functions $h(t)$ and $H(t)$ must be established. Teodorovic [155] surveyed passengers traveling on the Belgrade-Zagreb route (Yugoslavia).

Passengers were asked the following question :

- If you were given the choice, what time would you choose to fly from Belgrade to Zagreb?

After tabulating the survey results, passenger flow as a function of the time of day on this route was :

$$h(t) = -0{,}0802 \cdot t^4 + 2{,}5827 \cdot t^3 - 25{,}9952 \cdot t^2 + 79{,}9252 \cdot t + 48{,}2199 \qquad (2.70)$$

i.e., the number of passengers wishing to travel up to time t :

$$H(t) = -0{,}016 \cdot t^5 + 0{,}6457 \cdot t^4 - 8{,}6651 \cdot t^3 + 39{,}9626 \cdot t^2 + 48{,}2199 \cdot t \qquad (2.71)$$

whereby time t=0 corresponds to 5:00 a.m. and time 9:00 p.m.

Using equation (2.61) it was possible to calculate average time delay per passenger without taking into consideration the effects of airplane departure times. Equations (2.66) and (2.69) enable the minimum time delay to be calculated between the actual and desired departure time and its comparison with the average time delay calculated using equation (2.61).

TABLE 2.2 Comparison of average time delay per passenger calculated using different formulas

Daily number of flights N	$s_d = \frac{1}{4} \cdot \frac{T}{N}$ *(min)*	*Minimum passenger delay (min)*
8	30	28,69
9	26,40	25,54
10	24	22,38
11	21,60	20,68
12	19,80	18,83
13	18	17,15
14	16,80	16,35
15	15,60	14,70

It is clear from Table 2.2 that average time delay gives a very good approximation of minimum time delay since the differences are negligible. So we can conclude that the minimum time delay per passenger that can be achieved with a known flight frequency can be very easily estimated. Real

TABLE 2.3 Time delays for different flight schedules

Aircraft departure time	*Average delay between actual and desired departure per passenger* (min)
$x_1=1$; $x_2=2$; $x_3=3$; $X_4=7$; $x_5=12$ $x_6=13$; $x_7=14$; $x_8=15$	36,71
$x_1=3$; $x_2=5$; $X_3=8$; $x_4=11$ $x_5=13$; $x_6=14$; $x_7=15$; $x_8=16$	45,86
$x_1=1$; $x_2=3$; $x_3=5$; $x_4=7$; $x_5=9$ $x_6=11$; $x_7=13$; $x_8=15$	30
$x_1=0,9215$; $x_2=2,2057$; $x_3=3,4867$ $x_4=4,9563$; $x_5=6,9615$; $x_6=10,6239$ $x_7=12,6087$; $x_8=14,2787$ (Schedule with least delay)	28,69
Value obtained based on expression (2.61)	30

time delay on a route also largely depends on the distribution of airplane departures during the day. In order to understand the effects of the departure time distribution on real time delay, the results given by equation (2.61) will be compared with those obtained for several arbitrarily chosen flight schedules. For flight frequency $N=8$ on the route with previously described characteristics, 3 arbitrarily chosen departure times are considered and corresponding time delays are calculated. The results obtained are presented in Table 2.3.

2.10. Determining Flight Frequency on a Route With Competition

Previous discussions treated flight frequency determination on a route that maximized the air carrier's profit, i.e. that transported the greatest number of passengers when there is only one carrier on the route. The logical continuation is to determine flight frequency on a route that maximizes carrier profit when there are two or more carriers. This problem was considered by Powell [121] who assumed that there were two carriers. As with Renard, [126] Gelerman and de Neufville, [48] de Neufville and King, [24] Powell also assumed that there is a relation ship between the market share MS_i of carrier i and the carrier's frequency share , N_i having this shape :

$$MS_i = \frac{N_i^{\alpha}}{N_1^{\alpha} + N_2^{\alpha}} \qquad i = 1,2 \tag{2.72}$$

where α is the parameter to be estimated.

Airline 1 will be hereafter called the "carrier" and Airline 2 will be the "competitor". The total number of passengers wanting to travel λ_0 is a random variable that Powell, in order to simplify the problem, assumed to be distributed according to the normal probability density function with mean μ and standard deviation σ . Since we pointed out earlier that the number of passengers on any flight is a random variable most often distributed by the normal probability density function, the assumption that the total number of passengers wanting to travel is also a random variable distributed by the normal probability density function is very acceptable.

We denote by λ_1 the number of passengers to be transported by Air Carrier 1. It is clear that :

$$\lambda_1 = MS_1 \cdot \lambda_0 \tag{2.73}$$

Since :

$$MS_1 = \frac{N_1^{\alpha}}{N_1^{\alpha} + N_2^{\alpha}} = \frac{1}{1 + N_1^{-\alpha} \cdot N_2^{\alpha}} \tag{2.74}$$

Then :

$$\lambda_1 = \frac{\lambda_0}{1+N_1^{-\alpha} \cdot N_2^{\alpha}} \tag{2.75}$$

when Carrier 1 attracts less passengers than the number of seats offered.

The total number of seats offered by Carrier 1 equals $n \cdot N_1$, where n is the number of seats in the plane and N_1 the flight frequency of Carrier 1. We can finally write that the number of passengers to be transported by Carrier 1 is:

$$\lambda_1 = \begin{cases} \dfrac{\lambda_0}{1+N_1^{-\alpha}\cdot N_2^{\alpha}} & \lambda_1 \leqslant n\,N_1 \\ n \cdot N_1 & otherwise \end{cases} \tag{2.76}$$

Since the total number of passengers wanting to travel is a random variable assumed to be distributed according to the normal probability density function, then total revenues and total profit are also random variables. The mean of total profit $M(P)$ **on the route is :**

$$M(P) = \int_0^{\infty} c\cdot\lambda_1\cdot f(\lambda_0)d\lambda_0 - N_1\cdot K(n,d) \tag{2.77}$$

where :

c · **average ticket price per passenger,**

$f(\lambda_0)$ - **probability density function of random variable** λ_0

$K(n,d)$- **total costs per flight for an airplane with** n **seats flying on a route with length** d **.**

Since :

$$P\left[\lambda_1 \leqslant n \cdot N_1\right] = P\left[\frac{\lambda_0}{1+N_1^{-\alpha}\cdot N_2^{\alpha}} \leqslant n \cdot N_1\right] =$$

$$= P\left[\lambda_0 \leqslant nN_1(1+N_1^{-\alpha}\ N_2^{\alpha})\right] \tag{2.78}$$

then :

$$M(P) = \int_0^{nN_1(1+N_1^{-\alpha}N_2^{\alpha})} c\lambda_1\ f(\lambda_0)d\lambda_0 + \int_{nN_1(1+N_1^{-\alpha}N_2^{\alpha})}^{\infty} c\lambda_1\ f(\lambda_0)d\lambda_0 - N_1\cdot K(n,d) \tag{2.79}$$

And :

$$M(P) = \int_{0}^{nN_1(1+N_1^{-\alpha}N_2^{\alpha})} \frac{c\lambda_0 \cdot f(\lambda_0)}{1+N_1^{-\alpha}\;N_2^{\alpha}}\, d\lambda_0 + \int_{nN_1(1+N_1^{-\alpha}N_2^{\alpha})}^{\infty} cnN_1 f(\lambda_0) d\lambda_0 - N_1 K(n,d) \quad (2.80)$$

From condition

$$\frac{d\left[M(P)\right]}{dn} = 0 \quad (2.81)$$

the number of seats in the airplane is obtained that maximizes the air carrier's profit. By finding the first derivation and setting it equal to zero we get the following:

$$F\left[N_1 \cdot n^*(1+N_1^{-\alpha}N_2^{\alpha})\right] = 1 - \frac{\frac{d\left[K(n,d)\right]}{dn}}{c} \quad (2.82)$$

Powell showed by calibrating the network of American airways that total costs per flight $K(n,d)$ can only be approximately expressed as a function of the number of seats in the airplane n and the distance flown d as follows :

$$K(n,d) = \$\ 5.6n + (0{,}46+0{,}0093n)d \quad (2.83)$$

It is certain that calibrating another transportation market would produce somewhat different values for the parameters in equation (2.83)

Using equations (2.82) and (2.83), we get that the number of seats in the airplane n^* that maximizes the carrier's profit must satisfy the following :

$$F\left[N_1\;\; n^*(1+N_1^{-\alpha}N_2^{\alpha})\right] = 1 - \frac{5{,}6 + 0{,}0093\ d}{c} \quad (2.84)$$

For a normal probability density function of random

variable V_0 (with parameters μ and σ), relation (2.84) reads :

$$\Phi\left[\frac{N_1 n^*(1+N_1^{-\alpha}N_2^{\alpha}) - \mu}{\sigma}\right] = 1 - \frac{5,6+0,0093\ d}{c} \qquad (2,85)$$

where $\Phi(\cdot)$ is the Laplace function.

Using equation (2.84) or (2.85), for different values of $\mu, \sigma, c, d, N_1, N_2$ and α , it is possible to calculate the number of seats in the airplane that maximizes the air carrier's profit. Mention should be made that equations (2.84) and (2.85) only indicate the approximate capacity of the aircraft to be flown, since different parameter values in these equations can provide just about any number of seats in the airplane while only those produced with specific capacities can actually be used.

From condition :

$$\frac{d\left[M(P)\right]}{dN_1} = 0 \qquad (2.86)$$

we get :

$$F\left[nN_1^*(1+N_1^{*-\alpha}N_2^{\alpha})\right] = 1 - \frac{K(n,d) - \dfrac{\alpha N_2 N_1^{*-(1+\alpha)}}{\left[1+N_1^{*-\alpha}N_2^{\alpha}\right]^2}}{cn} \qquad (2.87)$$

where :

N_1^* - the flight frequency that maximizes the profit of Carrier 1 on the route.

A link is established in equation (2.87) between the capacity of the airplane n, flight frequency N_1^* that maximizes the carrier's profit, the flight frequency of competitive carrier N_2, average air fare c, and total costs per flight $K(n,d)$ for the case of stochastic transportation demand whose characteristics are stated through the form of the cumulative density function $F[\cdot]$. Using equation (2.87), a decimal number can also be obtained as the calculation result, while the flight frequency value must be a whole number. But in spite of this remark, the model encompasses all important factors that influence flight frequency and when applied considerably facilitates reaching appropriate decisions on how to organize air transportation on a route where the carrier has competition.

2.11. Simultaneously Determining Departure Times and Flight Frequency on a Route

Flight frequencies are normally determined on the branches of a transportation network before departure times are determined on individual routes, i.e. before designing the flight schedule. This means that using earlier determined flight frequencies for the established objective function, while observing certain constraints, airplane departure times are determined on all routes of the transportation network and aircraft are routed. The following objective functions are most often in operation (Simpson) [146]:

- Minimizing the fleet size,
- Minimizing operating costs,
- Maximizing the carrier's profit,
- Maximizing total benefits to the carrier and society.

Constraints in the model most often include the number of aircraft in the fleet, the different types of aircraft in the fleet, flights with several stopovers, minimum or maximum daily frequencies, the location of the base where aircraft must "sleep", daily utilization of aircraft, airport operating hours, technical maintenance service requirements in terms of the maximum number of days that an aircraft can spend flying without returning to the technical base, etc. For most of the models that deal with determining departure times and routing aircraft on the network, flight frequency on individual routes are the input data. Dispatch models are an exception to this in which flight frequency is determined simultaneously with departure times during the day on a route. The basic input data for dispatch models is the functional dependence of the passenger flow over time during the day $h(t)$ on the route. This dependence is obtained by recording the arrival of passengers for an air transportation system without ticket reservations or based on surveying the passengers and establishing requested travel times when this concerns an air transportation system with the possibility of reserving a seat on the airplane.

Simpson [146] considered the problem of simultaneously determining aircraft departure times and flight frequency when the shape of passenger flow $h(t)$ is known, to the effect that he focussed on minimizing carrier operating costs and passenger waiting time. Simpson proposed dynamic programming as a suitable technique to solve this problem.

Let us note the route between city A and city B with known function $h_{AB}(t)$ (Figure 2.11). In order to simplify things, the passenger flow is assumed to be uniform.

FIGURE 2.11 Passenger flow

Let passenger transportation demand be expressed during time period (O,T) . We assume that this is an air transportation system without ticket reservations so that each passenger who cannot find a vacant seat on the airplane with the first departure time must try to get a vacant seat on a following flight. Let the capacity of the airplane equal n . We divide time interval $(0,T)$ into m small intervals of width Δt so that :

$$m\Delta t = T \tag{2.88}$$

We also assume that one passenger appears during each of the small intervals Δt . So we have :

$$h_{AB}(t) = \begin{cases} \frac{1}{\Delta t} & 0 \leq t \leq T \\ 0 , & otherwise \end{cases} \tag{2.89}$$

The total number of passengers $H_{AB}(T)$ that want to travel from city A to city B during time interval $(0,T)$ equals:

$$H_{AB}(T) = \int_0^T h_{AB}(t)dt = \frac{T}{\Delta t} = m \tag{2.90}$$

Figure 2.12 shows the network in which the dispatch problem is being solved. Stages in this network represent individual moments of time, and the states represent the number of passengers waiting at times $0, \Delta t, 2\Delta t, \ldots, m\Delta t$.

As can be seen on Figure 2.12, at some time $i\Delta t$ $(i = 1,2,3,\ldots,m-1)$ the number of waiting passengers z_i can equal 0,1,2,..., or i passengers. At time $m\Delta t = T$ the number of waiting passengers must equal zero since all passengers wanting to travel during time interval $(0,T)$ must be transported by the end of time $m\Delta t = T$

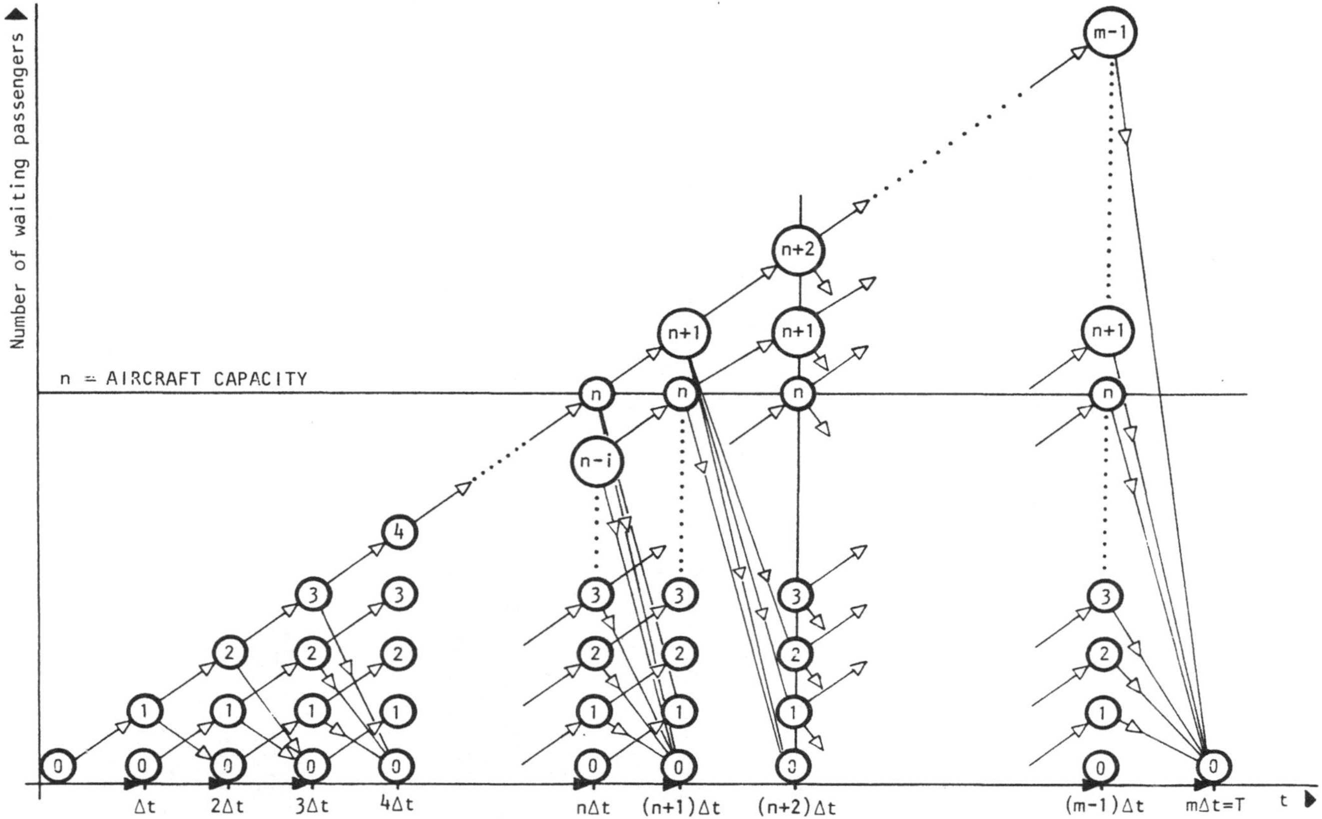

FIGURE 2.12 Network where the dispatch problem is solved

It was mentioned earlier that flight frequency and departure time will be determined so that operating expenses and passenger waiting time are minimized. Let us consider time $(n+1)\Delta t$ with the number of passengers waiting at this moment equal to $n+1$. If the number of passengers waiting at time $(n+2)\Delta t$ is $(n+2)$, this means that there was no departure at time $(n+2)\Delta t$, either. Here we take the length of the branch going from node $(n+1)$ at time $(n+1)\Delta t$ to node $(n+2)$ at time $(n+2)\Delta t$ to be the waiting time cost of the passenger who arrived during time interval $[(n+1)\Delta t, (n+2)\Delta t]$. We take the length of the branch going from node $(n+1)$ at time $(n+1)\Delta t$ to node 2 at time $(n+2)\Delta t$ to be the sum of the operating costs for transporting n passengers and the waiting time cost of the passenger that arrived during time interval $[(n+1)\Delta t, (n+2)\Delta t]$.

We note branch (z_{i-1}, Z_i) linking passenger number z_{i-1} waiting at time $(i-1)\Delta t$ with passenger number z_i waiting at time $i \cdot \Delta t$. We denote by $R_i(z_{i-1}, z_i)$ the length of branch (z_{i-1}, z_i). The branch length stands for the sum of the operating costs and passenger waiting time. (This means converting waiting time into monetary units beforehand). We denote by $f_i(z_i)$ the length of the shortest path to node z_i located in stage i. Since $i=0$ is the starting stage, then $f_0(0) = 0$.

The dynamic programming equations read :

$$f_0(0) = 0, \tag{2.91}$$

$$f_i(z_i) = min\{f_{i-1}(z_{i-1})+R_i(z_{i-1},z_i)\} \tag{2.92}$$

$$i = 1,2,\ldots,m$$

Equations (2.91) and (2.92) let us determine flight frequency and aircraft departure times so that the sum of carrier operating costs and passenger waiting time is minimized.

Flight frequency is most often determined simultaneously with aircraft daily departure times for shuttle services. The basic characteristics of this type of air transportation organization include equal time intervals between aircraft departures and the impossibility of reserving a seat in advance. Since the passenger in this case has lost the possibility of reserving a seat in advance, some sort of guarantee must be provided in terms of waiting time. For example, the passenger can be guaranteed a seat on the first flight, or he can be guaranteed a seat on the next flight with a certain probability, or he can be guaranteed a seat for sure on the next flight if he did not find a vacant seat on the departure he wanted, etc. So, in spite of the fact that there is no seat reservation, the passenger can still be provided with certain guarantees by the carrier concerning his departure waiting time. The guarantee given to passengers depends on the average daily number of passengers, the shape of passenger flow, the desired average load factor, the number of planes in operation, total operating costs, average passenger waiting time, the probability of finding a vacant seat on the first departing airplane, etc. There is a strong relationship between the values of these variables and flight frequency and the time interval between two consecutive departures. Flight frequency and the time interval between two consecutive departures chosen by the carrier are conditioned by both the planned economic effects and by the quality of transportation services that the carrier plans to provide. So in order to determine flight frequency for a shuttle, calculations must be made for several different daily flight frequencies, i.e. time intervals between consecutive departures, concerning needed transportation capacities, total operating costs, average passenger waiting time, the probability of finding a vacant seat on the first departing plane, etc. Then the flight frequency that produces the best effects is chosen. Since the choice of a shuttle's flight frequency depends on several conflicting criteria, multicriteria decision making methods are very convenient in this case. Teodorovic [159] applied the TOPSIS method (Technique for Order Preference by Simularity to Ideal Solution) of Hwang and Yoon [66] when ranking several alternative shuttles that differed by the size of the interval between two consecutive departures, i.e. by flight frequency, and the probability of finding a vacant seat in the first departing aircraft.

Let us calculate the parameters that are used to choose flight frequency (i.e. the time interval between two consecutive departures) when at least p percentage of the passengers can find a vacant seat on the first departing airplane and when every passenger is guaranteed that he will find a vacant seat on the next aircraft if he did not find one on the first departing plane.

Let passenger demand for transportation between two cities be expressed during time interval $(0,T)$. We denote by $x, 2x, \ldots, zx$ the departure time of the first, second,... aircraft on the route under observation, where :

$$zx = T \tag{2.95}$$

A passenger appearing during time interval $(i-1)x$ to $ix, i=1,2,\ldots,z$ will be able to find a vacant seat on the airplane departing at time ix with a probability at least equal to p. If the passenger does not find a vacant seat on the airplane departing at time ix, he will find a seat with a probability equal to 1 on the airplane departing at time $(i+1)x$. This means that every passenger is guaranteed with a probability that at least equals p that he will wait for departure less than x, and with a probability of 1 that he will wait for departure less than $2x$. We denote by t_w passenger waiting time for departure. Thus :

$$P(t_w < x) \geqslant p \tag{2.94}$$

$$P(t_w < 2x) = 1 \tag{2.95}$$

We denote by $k_1, k_2, \ldots, k_z$ the least number of seats that must be offered to passengers at times $x, 2x, \ldots, zx$. In order to satisfy equations (2.94) and (2.95), Teodorovic[159] showed that the following must be fulfilled :

$$k_1 = pH(x) \tag{2.96}$$

$$k_i = \{H(ix) - H[(i-1)x]\} + (1-p)\{H[(i-1)x] - H[(i-2)x]\} \tag{2.97}$$

$$i = 2, 3, \ldots, z \qquad H(0) = 0$$

where :

$H(t)$ - the total number of passengers wanting to travel up to time t .

The number of aircraft f_i departing at times $x, 2x, \ldots, zx$ equals :

$$f_i = \left[\frac{k_i}{n}\right] + 1 \quad , \; i = 1, 2, \ldots, z \tag{2.98}$$

where :

n - the number of seats in the plane in operation,

$[y]$ - the integer of number y .

The actual number of seats offered at times $x, 2x, \ldots, zx$ equals nf_i, $i = 1, 2, \ldots, z$. For known values of f_i it is easy to calculate the number of aircraft needed to operate a shuttle service between two cities. A very good method to calculate the least number of aircraft needed for a given airline schedule was given by A.M. Lof [94] and Levin [90]. The papers of Gillett and Miller [56] and Fisher and Jaikumar give a simple heuristic method for calculating the needed number of vehicles.

Figure 2.13 present a space-time diagram in which the nodes are flights to be flown.

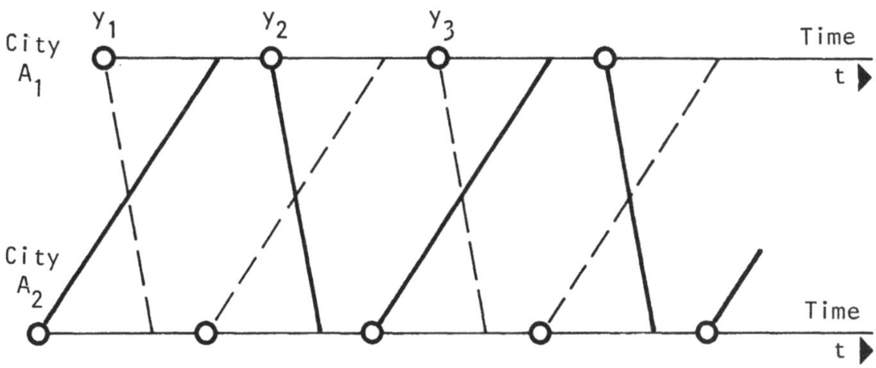

FIGURE 2.13 Space-time diagram

Nodes are denoted respectively by $y_1, y_2, y_3, \ldots$. A simple visual inspection of the space-time diagram after joining corresponding nodes (taking block times and ground service time into consideration) enables the easy determination of the route of each aircraft in the fleet and the total number of aircraft needed. Figure 2.14 shows the number of passengers $H(t)$ wanting to travel before time t and the number of passengers $D(t)$ that departed before time t .

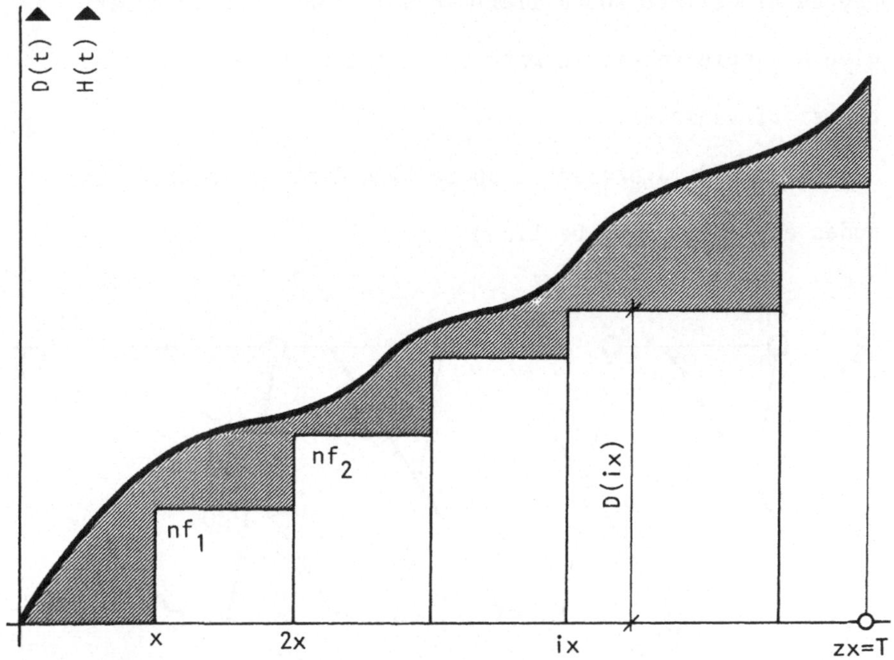

FIGURE 2.14 The number of passengers wanting to depart and the number of departed passengers as a function of time

The total wait of all passengers W wanting to travel during time interval $(0,T)$ (shown by the shaded area in Figure 2.14) equals :

$$W = \int_{0}^{T} H(t)dt - x \sum_{i=1}^{z-1} D(ix) \tag{2.99}$$

where :

$$D(ix) = min \left\{ \sum_{j=1}^{i} nf_j, \; H(ix) \right\} \tag{2.100}$$

Average waiting time $\overline{t}_w$ per passenger is :

$$\overline{t}_w = \frac{W}{H(t)} \tag{2.101}$$

The number of passengers Q who cannot find a vacant seat on the first departing flight equals :

$$Q = \sum_{i=1}^{z} \left[H(ix) - D(ix) \right] \tag{2.102}$$

The probability of finding a vacant seat q on the first departing flight equals :

$$q = 1 - \frac{Q}{H(T)} \tag{2.103}$$

where $q \geqslant p$.

Based on the known number of flights during the day, it is easy to calculate the shuttle service's total costs. This, plus total number of engaged aircraft, average passenger waiting time and the probability of finding a vacant seat on the first departing plane, if calculated for several different time intervals between two departures, provides the best alternative for the shuttle service. It is clear that the best alternative, i.e. daily frequency, between two cities highly depends on the importance that the decision maker gives to the variables that make up the calculation.

Shuttle services have been put into widespread operation

in recent years. Theoretical models and their practical implementation on actual routes have been primarily the work of airline company operations research departments. Jose Luis Gasco [44] presented the partial results of one such research project. His paper treated the experience of IBERIA, the Spanish national airline company, regarding the introduction of a shuttle between Madrid - Barcelona in 1974. During 1972 and 1973 the average daily number of passengers was 1250 on the Madrid - Barcelona route and 1280 on the Barcelona - Madrid route. The average load factor for a classical fixed schedule was 71%. IBERIA was primarily interested in increasing the quality of services. To this effect, it was established that 77.6% of the passengers on this route were on business trips. Businessmen did not cancel reservations and it was also shown that if they finished their work early they would go to the airport, get on a waiting list and try to leave on the first available flight. It was also noticed that many businessmen came to the airport with open tickets and no reserved seats, since their frequent trips made them aware of the possibilities of finding vacant seats on certain flights. IBERIA provided passengers the following guarantees through the new flight schedule :

1) A 100% guarantee of departing on the desired flight if the flight can be reserved in advance and if the seat is also reserved in advance. In other words,

there is a probability of 1 that the passenger will leave on a flight if the reservation is made on time.

2) A 90% guarantee that the passenger will leave on his desired flight if the flight has no seat reservations.

3) A 100% guarantee that the passenger will leave on the next flight if he did not find a seat on the flight of his choice.

Jose Luis Gasco developed a simulation model of the flight schedule. The simplicity of the simulation model makes it a very suitable technique to establish the relationship between flight frequency and shuttle service departure times on the one hand, and the number of needed aircraft, total transportation costs, average passenger waiting time, and the probability of finding a vacant seat on the first departing plane, on the other. As already mentioned, multicriteria decision making methods (Hwang and Joon)[66] are very convenient when ranking different possible shuttle service alternatives.

2.12 Determining Flight Frequency on a Network that Provides the Highest Quality of Transportation Services with Existing Transportation Capacities

Previous presentations discussed only the problems of determining flight frequency on one route. This section will treat the simultaneous determination of flight frequency on all branches of the transportation network. The problem of determining flight frequency on all branches of the transportation network is far more complex than the problem of determining the frequency on one route, since passenger demand between all pairs of cities in the network must be optimally matched to available transportation facilities. This is done by designing the network, choosing the type of aircraft to operate on the routes and determining flight frequency all at the same time. To this effect, flight frequency on the network branches, total economic effects realized by the carrier and the quality of transportation services all depend on the capacity of the aircraft in operation. Gordon and de Neufville [58] were among the first to study the problem of designing an air network and determining flight frequency on all of its routes. They used schedule delay as a measure of the quality of transportation services provided to passengers, starting with the assumption that total schedule delay D can be expressed as :

$$D = a \sum_{i} \frac{\lambda_i}{(1-\eta_{p_i})\, N_i} \tag{2.104}$$

where :

a - a constant proportionality,

N_i - flight frequency on link i ,

λ_i - the number of passengers on link i ,

η_{p_i} - the average load factor on link i .

Gordon and de Neufville assumed that delays on a link decrease with a rise in frequency, i.e. with a decline in the load factor.

It is clear that a rise in the flight frequency on a link decreases total passenger delay. This rise, or decrease, can go up to a certain limit that is set by the size and structure of the carrier's fleet. The air carrier can offer passengers a fixed number of seat-hours during a set period of time. This means that the number of available seat-hours S must satisfy the following :

$$\sum_i N_i \cdot n \cdot t_{B_i} \leqslant S \tag{2.105}$$

where :

S - the number of available seat-hours that the carrier can offer passengers during a fixed period on the network,

n - the aircraft capacity, i.e. the number of seats in the aircraft in operation,

t_{B_i} - block time on the i-th route,

N_i - flight frequency on the i-th route.

Since the average load factor on the i-th route equals:

$$\eta_{p_i} = \frac{\lambda_i}{N_i \cdot n} \tag{2.106}$$

we get by substituting (2.106) into (2.104) :

$$D = a \cdot n \sum_i \frac{\eta_{p_i}}{1-\eta_{p_i}} \qquad (2.107)$$

And by substituting (2.106) into (2.105) we get :

$$\sum_i \frac{\lambda_i t_{B_i}}{\eta_{p_i}} \leqslant S \qquad (2.108)$$

In order for the air carrier to provide passengers with the highest possible transportation service with available capacities, total schedule delay must be minimized while satisfying equation (2.108), i.e. :

$$D = a\, n \sum_i \frac{\eta_{p_i}}{1-\eta_{p_i}} \longrightarrow min \qquad (2.109)$$

$$\sum_i \frac{\lambda_i t_{B_i}}{\eta_{p_i}} \leqslant S \qquad (2.110)$$

By introducing Lagrangian function $F(\eta_{p_i}, \alpha)$ and the Lagrange multiplier α we have :

$$F(\eta_{p_i}, \alpha) = an \sum_i \frac{\eta_{p_i}}{1-\eta_{p_i}} + \alpha \left(\sum_i \frac{\lambda_i t_{B_i}}{\eta_{p_i}} - S \right) \rightarrow min \qquad (2.111)$$

From condition :

$$\frac{\partial F}{\partial \eta_{p_i}} = 0 \qquad i \qquad \frac{\partial F}{\partial \alpha} = 0 \qquad (2.112)$$

we get :

$$\alpha = \frac{an\left[\sum_i \sqrt{\lambda_i t_{B_i}}\right]^2}{\left[S - \sum_i \lambda_i t_{B_i}\right]^2} \qquad (2.113)$$

i.e. :

$$N_i = \frac{\lambda_i}{n} + \frac{S - \sum_i \lambda_i t_{B_i}}{n \sum_i \sqrt{\lambda_i t_{B_i}}} \sqrt{\frac{\lambda_i}{t_{B_i}}} \qquad (2.114)$$

Equation (2.114) gives the values of the optimal flight frequencies on each route in the network so that total passenger schedule delay is minimized.

Flight frequencies N_i are obtained as a function of the average number of passengers on the routes (λ_i), the possibilities of the carrier (n, S) and the characteristics of each route given in the network (t_{B_i}). Based on equation (2.114), the average load factor on the i-th route η_{p_i} equals:

$$\eta_{p_i} = \frac{\sqrt{\lambda_i t_{B_i}} \cdot \sum_i \sqrt{\lambda_i t_{B_i}}}{\sqrt{\lambda_i t_{B_i}} \sum_i \sqrt{\lambda_i t_{B_i}} + (S - \sum_i \lambda_i t_{B_i})} \qquad (2.115)$$

Inspired by the work of Gordon and de Neufville, Teodorovic [162] applied a similar methodological procedure to determine flight frequencies. He started with the assumption that average schedule delay per passenger on the i-th route s_{d_i} equals :

$$s_{d_i} = \frac{T}{4N_i} \tag{2.116}$$

where T is the time period expressing passenger transportation demand.

Equation (2.116) is actually equation (2.61). When analyzing the effects of flight frequency on average delay, it was shown that equation (2.61), i.e. (2.116), can be used to give an approximation of average schedule delay.

Total delay for all passengers in the network equals :

$$W = \sum_{i=1}^{p} s_{d_i} \cdot \lambda_i \tag{2.117}$$

where p is the number of routes in the network.

It is easily shown that :

$$W = \frac{T}{4} \sum_{i=1}^{p} \frac{\lambda_i}{N_i} \tag{2.118}$$

Average delay per passenger on the entire network is :

$$w = \frac{T}{4 \sum_{i=1}^{p} \lambda_i} \cdot \sum_{i=1}^{p} \frac{\lambda_i}{N_i} \tag{2.119}$$

As in the previous case, flight frequency is determined with the stipulation that the number of seat-hours realized is equal to or less than the number of available seat-hours S . The number of available seat-hours S is a function of the number of aircraft in the fleet, aircraft capacity, the flight schedule, maintenance system, etc. This number can be established using the following equation :

$$\frac{\sum_{i=1}^{p} \lambda_i t_{B_i}}{S} = m \qquad (2.120)$$

where m is the utilization rate of available seat-hours.

Parameter m depends on the air carrier's policy as to whether he is ready to earn less income in order to improve the quality of his transportation services.

With the assumption that average delay per passenger is defined by equation (2.116), flight frequencies $N_i (i=1,2,\ldots,p)$ on each route are obtained as the outcome of the following problem.

$$W = \frac{T}{4} \sum_{i=1}^{p} \frac{\lambda_i}{N_i} \rightarrow min$$

$$\sum_{i=1}^{p} N_i n t_{B_i} \leqslant S \qquad (2.121)$$

By introducing Lagrangian function $F(N_i, \beta)$ and Lagrange multiplier β we have :

$$F(N_i, \beta) = \frac{T}{4} \sum_{i=1}^{p} \frac{\lambda_i}{N_i} + \beta\left(\sum_{i=1}^{p} N_i n t_{B_i} - S\right) \rightarrow min \qquad (2.122)$$

From condition :

$$\frac{\partial F}{\partial \beta} \quad i \quad \frac{\partial F}{\partial N_i} = 0 \quad , \ i = 1,2,\ldots,p \qquad (2.123)$$

we get :

$$N_i = \frac{S\sqrt{\lambda_i}}{n\sqrt{t_{B_i}} \sum_{i=1}^{p} \sqrt{\lambda_i t_{B_i}}} \quad , \quad i = 1,2,..,p \qquad (2.124)$$

Average load factor on the i-th route in this case is defined by the expression :

$$\eta_{p_i} = \frac{\sqrt{\lambda_i t_{B_i}} \sum_{i=1}^{p} \sqrt{\lambda_i t_{B_i}}}{S} \qquad (2.125)$$

When all aircraft in the fleet have the same capacity, the following equation is satisfied :

$$S = A \cdot n \cdot u \qquad (2.126)$$

where :

A - the number of aircraft in the fleet

u - the maximum number of daily aircraft utilization hours during time interval $(0,T)$

If the values of S and u are known, it is very easy to calculate the needed number of aircraft in the fleet, althought this is only an approximation. The exact number of aircraft needed can only be obtained after designing the airline schedule. Also, the exact rate of daily aircraft utilization is only obtained after designing the flight schedule.

By substituting (2.126) into (2.124) we get :

$$N_i = \frac{A\ u\ \sqrt{\lambda_i}}{\sqrt{t_{B_i}}\ \sum_{i=1}^{p} \sqrt{\lambda_i\ t_{B_i}}} \quad , \ i = 1,2,\ldots,p \quad (2.127)$$

Equations (2.124) and (2.127) define flight frequencies that give the passengers the highest quality transportation service, i.e. that minimize total passenger delay using available transportation capacities. It is interesting to compare the flight frequencies obtained from equation (2.114) with that from equation (2.124).

Table 2.4 presents flight frequencies calculated using equations (2.114) and (2.124). The Table also presents the characteristics of the routes for which flight frequencies are estimated. It is assumed that the aircraft in opertion has a capacity $n = 100$ seats and that the carrier is able to offer passengers $S = 10{,}000$ seat-hours a day.

It can be seen from the Table that both formulas give approximately the same flight frequencies. Equation (2.114) gives a somewhat greater flight frequency than equation (2.124) on routes with more passengers. For routes with less passengers, equation (2.124) gives a slightly greater flight frequency.

Equations (2.114) and (2.124) establish the link between air carrier characteristics (n and S), the characteristics of the operating network $(\lambda_i, t_{B_i},\ i=1,2,\ldots,p)$ and

TABLE 2.4 Flight frequencies calculated by different formulas

Route i	Average daily No. of passengers in in one direction λ_i	Block time t_{B_i} (h)	Equation (2.114) $N_i = \frac{\lambda_i}{n} + \frac{S - \sum_i \lambda_i t_{B_i}}{n \sum_i \sqrt{\lambda_i t_{B_i}}} \sqrt{\frac{\lambda_i}{t_{B_i}}}$	Equation (2.124) $N_i = \frac{S\sqrt{\lambda_i}}{n\sqrt{t_{B_i}} \sum_i \sqrt{\lambda_i t_{B_i}}}$
1	250	1,5	4,46 ∿ 4	5,22 ∿ 5
2	300	1,4	5,22 ∿ 5	5,92 ∿ 6
3	350	1,6	5,75 ∿ 6	5,98 ∿ 6
4	400	1,2	6,77 ∿ 7	7,38 ∿ 7
5	450	1,1	7,57 ∿ 8	8,17 ∿ 8
6	500	1,7	7,61 ∿ 8	6,93 ∿ 7
7	550	1,3	8,63 ∿ 9	8,31 ∿ 8
8	600	1,1	9,55 ∿ 10	9,40 ∿ 9
9	650	1,3	9,90 ∿ 10	9,04 ∿ 9
10	700	1,2	10,67 ∿ 11	9,76 ∿ 10

flight frequencies $(N_i,\ i=1,2,\ldots,p)$. Flight frequencies are calculated so that passengers are offered the highest possible service quality considering available transportation capacities. Service quality is measured by the amount of total passenger schedule delay. A universal measure of service quality is given by the total delay of all passengers representing the sum of the total schedule delay and the total stochastic delay. A similar methodological procedure can be used to calculate flight frequencies on a network so that total passenger delay is minimized using available transportation facilities.

2.13 Determining Flight Frequencies on a Network that Minimize the Carrier's Operating Costs

The previous section dealt with determining flight frequencies that provides passengers the highest quality service with existing transportation capacities. From the passenger's viewpoint this approach is completely warranted. On the other hand, airlines, which in most countries do not receive financial assistance from the government, are compelled to focus on maximizing their profits, i.e. minimizing operating costs while considering a certain service quality. A logical result of this situation is the problem of determining flight frequencies on the network that minimizes air carrier costs. This problem was considered by Carter and Morlok [19] who determined flight frequencies that minimizes the carrier's direct operating costs on a network with both direct flights and flights with stopovers.

Flight frequency (direct flights) between city i and city j is denoted by N_{ij} (flight frequency in one direction). When there are flights between the two cities with one or more stopovers, then the aircraft flying between them can take different paths. How these paths are determined will be discussed later on. We denote by r and s the cities that have flights with one or more stopovers between them. Flight frequency in one direction between city r and city s

along path p is denoted by N_{rsp}. The basic input to calculate flight frequencies is the estimated daily number of passengers in one direction λ_{rs} between city r and city s. The daily number of passengers between city r and city s traveling along path p is denoted by λ_{rsp}. The least possible number of daily flights between city r and city s is denoted by L_{rs}. In these models this variable is a measure of the lowest quality of transportation service offered to passengers traveling between cities r and s. Thus, when determining flight frequencies that minimizes air carrier costs, a certain amount of consideration is given to the quality of the transportation services offered. Variables L_{rs} are also inputs given in advance. The flight frequency that minimizes the air carrier's direct operating costs is obtained by solving the following problem :

$$\sum_{i,j} c_{ij} N_{ij} \rightarrow min \qquad (2.138)$$

where :

$$C N_{ij} - \sum_{rsp \in ij} \lambda_{rsp} \geq 0, \; \forall \; (i,j) \qquad (2.139)$$

$$\sum_{p} \lambda_{rsp} = \lambda_{rs}, \; \forall \; (r,s) \qquad (2.140)$$

$$N_{ij} - N_{rsp} \geq 0, \; \forall (i,j) \quad for \quad rsp \in ij \qquad (2.141)$$

$$\sum_{p} N_{rsp} \geq L_{rs}, \; \forall (r,s) \qquad (2.142)$$

$$all \; N_{ij}, \; N_{rsp}, \; \lambda_{rsp} \geq 0 \qquad (2.143)$$

where all N_{ij}, N_{rsp} are integers where :

N_{ij} - daily flight frequency between cities i and j (direct flights),

c_{ij} - the cost of one flight from city i to city j ,

λ_{rsp} - the daily number of passengers traveling from city r to city s along path p ,

λ_{rs} - the total number of passengers between cities r and s ,

C - the effective capacity of the aircraft obtained by multiplying the number of seats in the aircraft n and the average load factor η_p ,

N_{rsp} - the daily flight frequency between city r and city s along path p .

We will explain in more detail the constraints defined by equations (2.139) and (2.143). In can be concluded from equation (2.139) that the daily offered number of seats on every link must be at least equal to the average daily number of passengers on all paths contained in the link under observation. Equation (2.140) defines the constraint that the sum of the passengers on individual paths leading from some city r to some city s equals the total number of passengers traveling between these two cities. This constraint must be satisfied for all pairs of cities between which flights are planned in the network. Flight frequency along any path must not be greater than the flight frequency along any of the branches that are part of the path. And this constraint defined by equation (2.141) refers to all paths in the network. Equation (2.142) determines the least allowed flight frequency between any two cities in the network. As already mentioned, this equation also has a built-in factor that considers the quality of transportation services. Equation (2.143) requires that all variables be non-negative. In addition, the flight frequencies obtained should be integers.

The above is a typical problem for integer programming. However, primarily due to the dimensions of the problem that could be expected in actual application, Carter and Morlok proposed that the problem be solved using linear programming whereby the constraint relating to whole numbers is disregarded. They illustrated the efficiency of this method on an example of using (V/STOL) aircraft in the Appalachian region of the USA. In order to estimate the number of trips between different pairs of cities in Appalachia, the authors used the abstract mode model and the multiple regression technique. They used a modified ATA method to calculate costs. When determining flight frequencies to minimize direct operating costs on more sizeable networks, a very large number of possible paths between two cities appears which further increases the dimensions of the problem to be solved. Carter and Morlok approached the task of reducing the number of possible paths in the following manner. They eliminated from further consideration all paths that were more than twice as long as the path that directly linked two cities with no stopovers. They also eliminated all paths with more than two stopovers and all paths containing a distance of less than 50 miles between two cities to the effect that there is no other city in the diameter equal to this. In this manner they considerably reduced the dimensions of the problem. The flight frequencies resulting from linear programming are rounded off to the nearest whole number to the effect that the value of the objective function is not essentially changed due to this rounding off.

In their model, Carter and Morlok assumed that all operating aircraft have the same capacity. Even if the aircraft in the problem to determine flight frequencies that minimizes costs or maximizes carrier profits are of different capacities, linear programming can still be used. In this case aircraft are assigned to different routes at the same time that flight frequencies are determined, which is why models that deal with this problem are called aircraft assignment models. Simpson[146] developed an assignment model that solves the following problem. For a given set of direct routes for which the carrier's frequency share is known (a route with competition) and for a given fleet composed of different types of aircraft, assign each type of aircraft to a route and determine their flight frequency so that profits are maximized, while also satisfying a set of constraints. These constraints concern the highest allowed load factor on any route, the availability of different types of aircraft, the smallest flight frequency on the routes, etc.

The problem of assigning aircraft from the fleet to the network is presented in the following sections.

2.14 Aircraft Assignment to the Network

Most international air carriers have different types of aircraft in their fleets, i.e. with different capacities and performances. Since the operating costs for different aircraft also differ, it is extremely important for the air carrier to correctly assign his aircraft to the network. In addition to the carrier's economic effects, aircraft assignment also has an important influence on the quality of service offered to passengers. For example, a larger capacity aircraft will fly on a route with a smaller flight frequency than a smaller aircraft, resulting in greater average schedule delay.

Aircraft assignment problems were considered by Maimaon and Peterson, [95] Konig, [80] Gerbracht [52] and Abara[1].

We will use a relatively simple example to explain the assignment of different types of aircraft to routes in the network. Let the air carrier fly with direct flights on a network with no competition and let there be K different types of aircraft in the fleet (carriers most often have from 2-5 types of aircraft in their fleets). Let us consider the following variables :

c_{ij} - the air fare for transporting a passenger on the route between city i and city j,

λ_{ij} - the average number of passengers on the route between city i and city j,

N_{ijk} - flight frequency between city i and city j with aircraft type k $(k = 1,2,\ldots,K)$

T_{ijk} - operating costs per flight by aircraft type k on the route between city i and city j

Profit D that the air carrier can earn equals :

$$D = \sum_{i,j} \sum_{k} \left[c_{ij}\lambda_{ij} - T_{ijk} \cdot N_{ijk}\right] \qquad (2.144)$$

We define the maximum allowed passenger load factor η_{ijk}^{**} for each type of aircraft in the fleet on a specific route. As indicated previously, an average load factor that is too small brings the air carrier poor economic results. On the other hand, an extremely high average load factor results in very poor quality transportation services, since in this case a very large number of passengers appears that did not succeed in getting a seat on the flight of their choice. Decisions concerning the specific value of variable η_{ijk}^{**} depend on the carrier's overall transportation policy. We denote by n_k the number of seats in aircraft type k. It is clear that the transportation capacities offered (offered number of seats) must be equal to or greater than the number of passengers requesting transportation on that route. In other words, the following must be fulfilled :

$$\sum_{k} \eta_{ijk}^{**} \cdot n_k \cdot N_{ijk} \geq \lambda_{ij}, \qquad \forall i,j \qquad (2.145)$$

We denote by u_k the maximum possible daily utilization of aircraft type k, and by A_k the total number of type k aircraft in the fleet. Variable u_k primarily depends on the structure of the network where aircraft type k operates, and on the technical maintenance system. We denote by t_{ijk} the flight time of aircraft type k on the route between city i and city j. Since total aircraft utilization must be equal to or less than the maximum possible utilization, this is :

$$\sum_{i,j} t_{ijk} \quad N_{ijk} \leq u_k \quad A_k, \qquad k=1,2,\ldots,K \quad (2.146)$$

During a certain period of time an equal number of take off and landing operations take place at each airport in the network. This means that :

$$\sum_{j} N_{ijk} = \sum_{j} N_{jik}, \qquad \forall i \qquad (2.147)$$

We denote by $N_{ij}^{(min)}$ the minimum flight frequency on the route between city i and city j . This is another variable that can serve as a measure of transportation quality. Flights between city i and city j can be made with one or several types of aircraft. However, the following must be satisfied regardless of the combination of aircraft types that fly between i and j :

$$\sum_{k} N_{ijk} \geqslant N_{ij}^{(min)}, \qquad \forall i,j \qquad (2.148)$$

In certain cases, airport administrations can restrict the carrier (to decrease overcrowdedness) to a maximum number of departures of all types of aircraft from a certain airport. This means that :

$$\sum_{k} \sum_{j} N_{ijk} \leqslant M_{i}^{(max)}, \qquad \forall i \qquad (2.149)$$

where $M_{i}^{(max)}$ is the maximum number of departures from airport i .

The carrier usually wants to find the aircraft route assignment that maximizes profit. Thus, the optimization problem reads :

$$D = \sum_{i,j} \sum_{k} \left[c_{ij}\lambda_{ij} - T_{ijk} \cdot N_{ijk}\right] \rightarrow max \qquad (2.150)$$

$$\sum_{k} \eta_{ijk}^{**} \cdot n_{k} \cdot N_{ijk} \geqslant \lambda_{ij}, \qquad \forall i,j \qquad (2.151)$$

$$\sum_{i,j} t_{ijk} \cdot N_{ijk} \leqslant u_k \cdot A_k, \quad k = 1,2,\ldots,K \tag{2.152}$$

$$\sum_{j} N_{ijk} = \sum_{j} N_{jik}, \quad \forall i \tag{2.153}$$

$$\sum_{k} N_{ijk} \geqslant N_{ij}^{(min)}, \quad \forall i,j \tag{2.154}$$

$$\sum_{k} \sum_{j} N_{ijk} \quad M_i^{(max)}, \quad \forall i \tag{2.155}$$

The above problem is solved using the linear programming method. Depending on how the problem is defined, several other constraints can appear or some of the present can be removed. The above-described model assumes that the carrier has a monopoly on the market. When assigning different types of aircraft to a network with competitive conditions, consideration must be taken of the functional dependence existing between the carrier's frequency share and his market share.

2.15 Planning Frequency and Traffic Assignment in the Network

Many trips between two nodes in an air transportation network can be made along several routes that lead from the origin to the destination. Many factors influence passengers' decision to travel on a certain route, including the most important of travel time, flight frequency and ticket price. Transportation can be by nonstop flights, direct flights with one or more stopovers or by indirect flights. The type of flight certainly has an important role in the passenger's choice of route. Since ticket prices are relatively unequal on different routes, it can be concluded that flight frequency along with travel time are of crucial importance to the assignment of traffic in the network, i.e. to passengers' choice of route. The problem of determining flight frequencies and traffic assignment was treated by Kanafani,[72] Kanafani and Ghobrial,[73] and Ghobrial.[54] Let us discuss the results obtained from these authors' research.

One of the main assumptions used as a starting point in flight frequency determination and traffic assignment models is that demand between different origins and destinations is stable as long as the transportation capacities offered are sufficient to satisfy transportation demand. On the other hand, it has already been mentioned that flight frequencies on routes are one of the most important factors affecting the assignment of passengers to routes. This means that the initial flight frequencies on routes will affect the number of passengers that choose to fly on certain routes. On the other hand, the number of passengers on certain routes will result in a change in flight frequencies if the previous frequencies did not satisfy passenger demand, while at the same time some other routes might have insufficiently utilized transportation capacities. New flight frequencies once again affect the passenger assignment, etc. It is clear that determining flight frequencies and passenger assignments are component parts of an iterative procedure that ends when flight frequency and the number of passengers on certain routes start to stabilize. This approach to determining flight frequencies originated with Kanafani.[72] By analyzing the network structure in short-haul traffic, Kanafani developed an iterative procedure that determines flight frequency on certain routes. This procedure is based on a multinomial Logit model of the form :

$$P(r,j) = \frac{e^{V(r,j)}}{\sum_{r \in R_j} e^{V(r,j)}} \qquad (2.156)$$

where :

$P(r,j)$- the percentage of passengers using route r when traveling between the j-th pair of cities,

R_j - the set of possible routes that can be used by the passenger when traveling between the j-th pair of cities,

$V(\cdot)$ - a linear choice function.

When calibrating the multinomial Logit model developed by Kanafani and Chang [71], we get :

$$V(r,j) = -0.0188 \cdot t_{rj} + 0.212N_{rj} \qquad (2.157)$$

where :

t_{rj} - travel time between the j-th pair of cities along route r,

N_{rj} - flight frequency between the j-th pair of cities along route r.

Calibration used data concerning 26 pairs of cities in the area in a 500 km radius around the city of Atlanta, Georgia (USA).

Using the iterative procedure, Kanafani alternatingly calculated the number of passengers, flight frequency and corresponding load factors on different routes. The iterative process was completed when average load factors were within acceptable limits. The lowest acceptable load factor on a link was the break even load factor provided in advance.

Ghobrial [54] similarly considered the problem of determining flight frequency and traffic assignment in the network. Noting that any branch in the network can be a component part of a larger number of routes between a fixed pair of cities,

Ghobrial started with the requirement that the following equations must be satisfied :

$$\sum_{i} \lambda_{ij}^{OD} = \sum_{m} \lambda_{jm}^{OD}, \qquad j \neq O,D, \ \forall O,D \tag{2.158}$$

$$\sum_{i} \lambda_{ij}^{OD} = \sum_{m} \lambda_{jm}^{OD} + D_{OD}, \qquad j=D, \ \forall \ O,D \tag{2.159}$$

$$\sum_{i} \lambda_{ij}^{OD} = \sum_{m} \lambda_{jm}^{OD} - D_{OD}, \qquad j=O, \ \forall O,D \tag{2.160}$$

where :

λ_{ij}^{OD} - the number of passengers traveling from O to D along branch (i,j) ,

D_{OD} - the total number of passenger traveling from O to D.

Equations (2.158), (2.159) and (2.160) indicate that the number of passengers that enter some node on the path from O to D equals the number of passengers leaving that node, i.e. that the number of passengers from all routes entering node D equals the total number of passengers traveling from O to D . Also, the flight frequency along any route k equals :

$$N_k = \min_{ij \in R_k} \{N_{ij}\} \tag{2.161}$$

where :

N_k - flight frequency along route k ,

N_{ij} - flight frequency along branch (i,j) ,

R_k - the set of branches contained in route k .

Ghobrial assumed that passenger assignment by available routes could be described using a multinomial Logit model of the following form :

$$\lambda_{r,j} = \frac{e^{V(r,j)}}{\sum_{r \in R_j} e^{V(r,j)}} \cdot D_j, \qquad \forall_j \tag{2.162}$$

$$\lambda^{(p,q)} = \sum_{r \in R^{(p,q)}} \lambda_{r,j} \tag{2.163}$$

where :

$\lambda_{r,j}$ - the number of passengers on route r traveling between the j-th pair of cities,

R_j - the set of possible routes that passengers can use when traveling between the j-th pair of cities,

$R^{(p,q)}$- the set of routes contained in branch (p,q) ,

$\lambda^{(p,q)}$- the number of passengers on branch (p,q) ,

$V(r,j)$- the linear function of the attributes of route r between the j-th pair of cities.

Function $V(r,j)$ represents a linear combination of the following attributes :

$$V(r,j) = a_1 \cdot T_{rj} + a_2 \cdot N_{rj} + a_3 \cdot c_{rj} + a_4 \cdot n_{rj} + a_5 \cdot G_{rj} \tag{2.164}$$

where :

T_{rj} - travel time between pairs of cities j along route r ,

N_{rj} - flight frequency between pairs of cities j along route r ,

c_{rj} - the price of an economy class ticket between pairs of cities j along route r ,

n_{rj} - a variable depending on the number of seats in the airplane (n),

G_{rj} - a variable depending on the type of flight,

$a_1, a_2, \ldots, a_5$ - parameters to be estimated.

Variables n_{rj} and G_{rj} are defined as follows :

$$n_{rj} = \begin{cases} 0, & n \leqslant 30 \\ .5, & 30 < n \leqslant 50 \\ 1, & n > 50 \end{cases} \qquad (2.165)$$

$$G_{rj} = \begin{cases} 0 & \text{for a nonstop flight} \\ .5 & \text{for a direct flight with one or more stopovers} \\ 1 & \text{for an indirect flights} \end{cases} \qquad (2.166)$$

We explained calibration of the multinomial Logit model n the first chapter. After calibrating the model, Ghobrial howed that the multinomial Logit model gives a very good escription of passenger assignment to available air transortation routes. With the known number of passengers on ndividual routes, flight frequency can be easily calculated ith the assumption that should there be competition, air arriers fly using the lowest acceptable load factor. Should here be a small number of passengers on a route so that the oad factor is below the limit, then no traffic is established along that branch and all routes of which the branch was a part. With these assumptions, Ghobrial proposed an iterative procedure to determine flight frequency and traffic assignment as follows :

Determine the set of possible routes between every pair of cities with air transportation established between them. In the next step determine the initial frequencies along individual branches so that the carrier achieves the lowest acceptable load factor. Calculate the number of passengers on different routes and branches using the initial flight frequencies and the multinomial Logit model for route choice. Calculate flight frequencies once more for the calculated number of passengers. With the new flight frequencies enter the multinomial Logit model once more, calculate the new number of passengers, etc. The iterative procedure ends when the flight frequency values and the number of passengers on individual routes start to stabilize.

The described iterative procedure determines flight frequencies and assigns passengers to routes. Flight frequencies obtained during the iterative procedure are corrected on the basis of passenger assignment obtained as a result of the previous flight frequencies. It should be underscored that corrections are made to the flight frequency on routes based on passengers' choice of routes.

3. AIRLINE SCHEDULE DESIGN MODELS

3.1. Factors that Influence Airline Schedule Design

Designing an airline schedule is one of the most complex tasks that air carriers encounter. A completed and adopted airline schedule is a working obligation for all those employed in the air carrier's services. Both the carrier's economic results and the quality of transportation offered to passengers depend on how the airline schedule is designed. When designing an airline schedule, consideration must be taken of passenger transportation demand on certain routes, the air carrier's economic interests and various operational constraints. Elaborating a network airline schedule is combinatorial by nature : of the large number of possible alternatives, those must be chosen that satisfy to the greatest extent the interests of the air carrier, the passengers and operational constraints. Passengers on the one hand are interested in the greatest possible flight frequency, departure times that are adapted to their desires, a high probability of finding a vacant seat on a particular flight, short waits to continue traveling at transfer points, few cancelled and late flights, etc. On the other hand, the carrier's interests are in a high load factor, high average annual aircraft utilization, decreased operating costs and increased profit. Carriers are interested in achieving a high annual utilization of aircraft that also have a high load factor, for if high average annual aircraft utilization is achieved with a low average load factor, the air carrier will have very poor economic results. It is also clear that a high average load factor with low aircraft utilization will give equally poor results. The air carrier is therefore interested in an airline schedule that results in a well "filled" airplane and good utilization of existing transportation capacities. Certain passenger requests regarding the airline schedule inevitably conflict with the carrier's requests. The airline schedule design must find the best possible way to reconcile these conflicting requirements.

Operational constraints complicate the coordination of various air carrier and passenger interests. The system of aircraft technical maintenance is the most important operational constraint that has a crucial influence on the choice of airline schedule alternatives. For reasons of safety, aircraft must have certain parts inspected or replaced at specific times at specific airports that are most often the air carrier's technical base. In addition, when greater technical work must be done, aircraft are removed from operations for a longer period of time which also has an important effect on the choice of airline schedule.

Meterological conditions prevailing at the airports in the network also act as a limiting factor when choosing a flight schedule. During the winter season, some regions often have airports that are closed to take off or landing operations which can result in many cancelled and late flights. In earlier deliberations we concluded that the air carrier should adjust his departures as much as possible to air passenger demand. However, if suitable attention during schedule designing is not given as well to the meteorological conditions prevailing in certain locations, coordinating departures to passenger demand will not bring suitable results since there will be numerous cancelled and late flights.

When elaborating the airline schedule, requirements from the crew planning service should also be studied in detail. Due to the fact that the flight crew primarily have licenses for one specific type of aircraft, care must be taken that there are enough pilots, copilots and flight engineers. Security reasons condition, and relevant laws and company regulations precisely define, the air crew's working hours, maximum allowed daily take offs and landing, needed rest, etc. Thus, when elaborating the airline schedule, attention must also be given to times during the day and airports where crew replacements will be made, the length of crew stays at certain airports while waiting to take on a job, the number of times when crew members travel as normal passengers in order to take on a job or to return to the base ("dead heading crews") and all other factors related to the work of the flight crew.

Aircraft can remain at airports for longer or shorter periods. The length of this detainment can also be considerably affected by fuel purchasing tactics. The price of fuel can vary remarkably between airports in the network both due to the tax policies of some countries and different suppliers. Since fuel costs are one of the carriers' largest outlays, they are obviously interested in purchasing fuel at certain airports in quantities that, while satisfying operational constraints concerning the allowed weight of the aircraft on take off and landing, also enable the carrier to minimize fuel costs. Should the carrier have developed a fuel purchasing strategy on the network, it is clear that this must be taken into consideration when designing the airline schedule.

Another constraint to remember when designing the airline schedule is the working hours of airports in the network. Some airports are open 24 hours a day. Most airports have prescribed working hours and only accept aircraft in emergency situations outside of these hours. Most international air carriers are interested in completing take off and landing operations within a fixed time interval during the day. A result of this desire is peak overload periods at airports that considerably hinder the work of airport services. As noted by Ashford,[6] with reference to London Heathrow, most airports in the world have introduced differential taxes for landing, accepting and dispatching aircraft during peak periods and outside of peak periods. These differential landing tariffs did not decrease the potential share of the number of operations during peak periods, but their very existence has resulted in their being built into the airline schedule model.

In long-haul traffic, different time zones present special operational constraints. Due to different time zones, a detailed study must be made of airport working hours, whether there is transportation between the airport and the city during different times of day, etc.

It is clear that designing an airline schedule is a very complex task since an adequate equilibrium must be found between passenger requests, carrier economic interests and possibilities and different operational constraints. As already mentioned in discussions about determining flight frequency, the number of passengers wanting to travel on a certain flight is a random variable most often distributed according to the normal probability distribution. The variability or stochastic aspect of passenger transportation demand considerably hinders airline schedule designing since fixed transportation facilities must be engaged for a long period of time in order to satisfy estimated passenger demand. Some air carriers tried to "control" passenger demand in order to get a better insight into it and thereby achieve a better load factor through the flight schedule design. As noted by Taneja,[153] the American airlines TWA tried to introduce a new concept on part of its network under the name demand scheduling. Passengers interested in flying on a certain day were to reserve at least three months in advance. A deposit of $20.00 was made at the same time as the reservation and was not returned if the reservation was cancelled. At the time the reservation was made, the passenger did not know the departure time, number of stopovers or the flight number. The air carrier provided passengers with flight information two months before departure. The air fare for this type of reservation was somewhat lower than the normal fare. Taneja indicates that in spite of the lower fare, this system did not produce expected results since a considerable share of the primarily business passengers were not interested in this type of transportation. This example makes it clear that "controlling" passenger transportation demand is far from simple and that the stochastic aspect of passenger demand must always be kept in mind when designing airline schedules.

A result of coordinating the carrier's interests and possibilities, passenger requests and the large number of operational constraints is that many departures can only be made during a fixed time interval during the day. Some schedule variations suit the passengers better while others are more satisfactory to the carrier's interests. The choice of schedule alternative is most certainly the result of the air carrier's overall transportation policy on a specific transportation market.

3.2 Types of Airline Schedules

The oldest and most frequent airline schedule is one that is prepared beforehand and is valid for a set period of time. Aircraft departure times for every day of the week are known for all routes on the network and are printed in publications intended for the transportation market. Passengers adjust their desired departure times to those given in advance in the airline schedule. Seats can be reserved in advance for this type of schedule and most passengers purchase their tickets in advance and reserve a seat well before departure time. The air carrier must keep to the announced schedule and flights must be made at the specified time regardless of the number of passengers in the plane. This type of schedule gives passengers complete information on the possibilities of making a trip. Such an airline schedule means that the air carrier has synchronized all of his other activities with the schedule.

Pre-announced airline schedules cover a longer period of time during which considerable changes often take place both in terms of passenger flows and their distribution over time. It follows from this that the basic drawback of the fixed airline schedule is its insufficient accommodation to actual passenger requests and needs. Simpson [146] comes to a similar conclusion, noting that the main defect of the fixed airline schedule is that it is "static" and not adjusted sufficiently to passenger needs which are constantly changing over time.

At the end of the 1960s, "dynamic scheduling" started to appear on certain routes in air transportation. The term "dynamic airline schedule" accurately describes the organization of air transportation without a fixed airline schedule. When there is no fixed airline schedule, air transportation can be organized in various ways. The common denominator of air transportation organized in this manner is the endeavor to adjust aircraft departure times as much as possible to passenger requests in order to provide the transportation market with high quality service. On routes with dynamic scheduling tickets and seats are most often not possible to reserve in advance. Different variations of dynamic scheduling are explained in greater detail in the chapter dealing with the simultaneous determination of aircraft departure times and flight frequency between two cities.

It was mentioned in previous discussions that some carriers tried to design airline schedules where the stochastic element in passenger demand was reduced to a minimum (demand scheduling). The basic goal of this type of schedule is to increase the average passenger load factor. In the future, various approaches to airline schedules should be expected that aim at bettering the air carrier's economic effects and/or improving the quality of transportation services.

3.3 A Classification of Airline Schedule Design Models

During the past four decades, numerous models have been developed to design airline schedules. Different operations research methods are utilized when modeling airline schedules depending on the approach to the design, the objective functions applied, and different operational constraints that must be considered. Due to the complex and combinatorial nature of the problem, modern computer techniques must be employed. As noted by Etschmaier and Mathaisel, [37] there are two approaches to airline schedule designing :

1. the direct approach to airline schedule design, and

2. the step-by-step approach to airline schedule design.

In the direct approach to airline schedule design, various heuristic procedures are employed. In recent years, the notion prevailed that an optimal airline schedule could not be obtained using the classical method of mathematical programming, primarily due to the large number of factors that could not be adequately quantified. The airline schedule has a direct effect on the work results of a large number of air carrier personnel. Therefore, the attempt to produce an optimal airline schedule meant an attempt to optimize the carrier's overall activities, which is certainly not possible. Due to the dimensions of the problem, both the direct approach and the step-by-step designing approach use computers. A large number of interactive computer programs to design airline schedules have been developed in recent years. Interactive computer programs are very convenient since they enable the airline schedule planner (who usually has considerable experience) to influence more or less the choice of schedule variation. Interactive computer programs are described in the works of Niederer,[110] Williamson,[170a] Smith and Kyle[147] and Deckwitz[22].

When using the step-by-step approach to airline schedule design, the first step determines fight frequency on individual routes. In the next step, possible departure times are determined taking into consideration both passenger requests and convenient times to arrive at important transfer points in order to allow passengers to continue their trip without much waiting. When departure times are established, the proposed airline schedule is tested in terms of operational constraints. When the departure times are adopted, aircraft are assigned to routes, i.e. each aircraft from the fleet is given specific flights to fly over a certain period of time after which the schedule is periodically repeated (usually one week). In the next step, since aircraft have been routed, an analysis is made of the schedule's flexibility, i.e. whether certain changes in the schedule might decrease the number of needed aircraft.

Although we noted earlier that the notion has prevailed in recent years that it is not possible to design an optimal airline schedule using classical mathematical programming methods, in many papers airline scheduling is still reduced to solving an optimization problem. Without considering certain factors that influence the airline schedule, an optimization problem can be defined in order to find the "optimal" airline schedule in relation to a definite objective function and constraints. An airline schedule obtained in this manner is certainly a solid basis for further corrections and improvements. As noted by Hyman and Gordon [67] and Simpson,[146] the objective function can represent maximizing the number of passenger-kilometers that earn a profit, maximizing the number of transported passengers, maximizing profit, maximizing the difference between total social benefits and costs, minimizing the number of aircraft in operation, minimizing direct and indirect carrier operating costs, etc. Constraints to the model can include the number of aircraft in the fleet, the fleet structure, flights with stopovers, minimum and maximum allowed frequencies on a route, maximum allowed number of daily operations at an airport, etc.

Depending on the problem modeled, the following types of airline schedule models are distinguished :

a) aircraft assignment models,

b) fleet planning models,

c) airline schedule design models for an isolated route,

d) airline schedule design models for a network, and

e) aircraft routing models.

We will give a short explanation of the above models.

Aircraft of different capacities and performances can be assigned to the network in different ways. The problem of finding the optimal aircraft assignment was discussed in the previous sections.

Fleet planning models solve the same problem but refer to a considerably longer period of time (5-10 years). These models use the forecasted passenger flow on routes to estimate the needed number of different types of aircraft and their assignment to the network.

Models to design airline schedules on an isolated route consider the problem of determining the departure time and number of departures on an isolated route. These models were explained in more detail in the previous chapter that dealt with flight frequency determination.

Designing airline schedules for the network comprises setting departure times on all routes in the network so that the given objective function is optimized while satisfying given constraints.

Aircraft routing models determine for known departure times on all routes the needed number of aircraft and their routes during a fixed time period (usually one week). This model classification should in no way be considered definitive. With the development of air transportation there has been constant changes and improvements to air schedule designing and operational control. New approaches and models to design airline schedules should therefore be expected in the future.

3.4. The Sequential Approach to Airline Schedule Design

Since the number of potential passengers on individual routes has been estimated, passenger requests have been studied in detail concerning departure times during the day on all routes in the network, and technical maintenance service requirements have also been thoroughly studied, since block-time* and flight frequency are known on the routes, as are service time and the meteorological situation at different airports, an approach can now be made to designing the airline schedule. Depending on the type of model, output can be departure times on different routes, an aircraft routing plan, the estimated number of passengers and estimated profit on individual flights.

In previous presentations it was noted that the functional dependence of passenger flow to the time of day characteristically shows two maximums or "peaks" - a morning and an evening. (the most frequent situation on short- and medium-haul routes). Figure 3.1. shows the characteristic functional dependence of flow intensity to the time of day on some route i .

The percentage number of transportation demands in a unit of time represents the percentage share of the number of passengers who want to travel at a certain time out of the total number of passengers wanting to travel during time interval $(0, T_i)$. Figure 3.1. presents departure times t and the time interval attraction of the flight departing at

* Flight block-time comprises the time that passes from motor start-up at the airport of origin to motor shut-down at the destination airport.

time t . Thus it is assumed that every flight has a corresponding attraction interval with a width of 2τ . In other words, it is assumed that all passengers who wanted to travel during time interval $(t-\tau, t+\tau)$ will choose the departure at time t . This approach to determining flight attraction intervals originated with Hyman and Gordon.[67] The assumption is certainly debatable that all flights have time interval attractions of equal length. As mentioned earlier, other approaches are found in the literature to determine the limit of attraction (Gagnon,[42] Soumis,[149] Teodorovic[155]). However, we

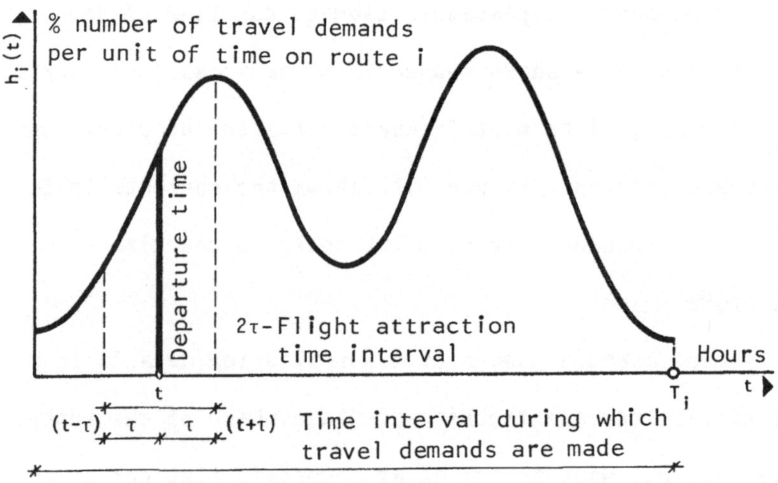

FIGURE 3.1 Percentage number of demands to travel per unit of time as a function of the time of day

will accept the assumption that all passengers who wish to travel during time interval $(t-\tau, \; t+\tau)$ will choose the departure at time t since this is one of the essential assumptions in the Hyman and Gordon model that we will present with certain modifications. We would mention from the outset that the width of the attraction time intervals is chosen arbitrarily and different airline schedules are obtained for different values of τ.

We denote by $s_i(t)$ the number of passengers that will choose the departure at time t on route i. It is clear that :

$$s_i(t) = \lambda_i \int_{t-\tau}^{t+\tau} h_i(t)dt \qquad (3.1)$$

where λ_i is the total number of passengers on route i that want to travel during time interval $(0, T_i)$.

We denote respectively by c_i and C_i the average price of a ticket on route i and the total costs per flight on route i for a specific type of aircraft. It is clear that revenues $P_i(t)$ and profit $D_i(t)$ that can be earned per flight on route i are a function of the chosen departure time t. This means :

$$P_i(t) = c_i \cdot \lambda_i \int_{t-\tau}^{t+\tau} h_i(t)dt \qquad (3.2)$$

i.e. :

$$D_i(t) = P_i(t) - C_i(t) = \left[c_i \lambda_i \int_{t-\tau}^{t+\tau} h_i(t)\,dt \right] - C_i \quad (3.3)$$

Depending on whether the revenue or profit maximizing criterium is chosen, one or the other variable is used when designing the airline schedule. Variable $D_i(t)$ should be used when several different types of aircraft enter into the airline schedule, since different types of aircraft result in different costs for the same route. Let us assume (in order to decrease the dimensions of the problem) that departures on all routes can only be made at certain times. In other words, we divide time interval T_i during which passenger transportation demands were made on route i into smaller time intervals with a width of Δt_i (Figure 3.2).

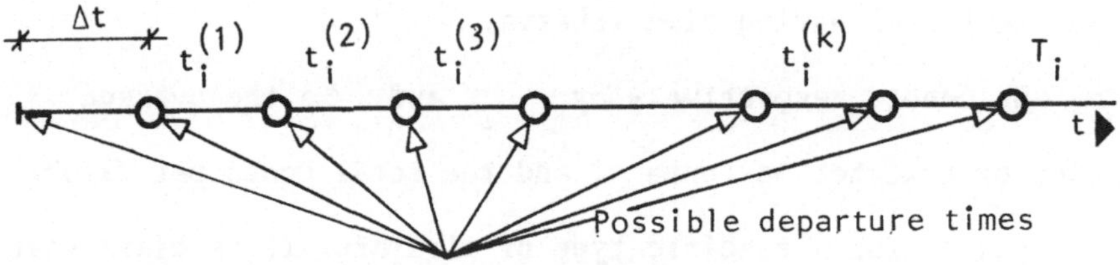

FIGURE 3.2 Possible departure times

We denote by $t_i^{(1)}, t_i^{(2)}, t_i^{(3)}, \ldots$ respectively the possible departure times on route i. Let there be m airports on the network with flights made between them, designated by order as $X_1, X_2, \ldots, X_m$. Let there be a total of t flights. This network is shown on Figure 3.3.

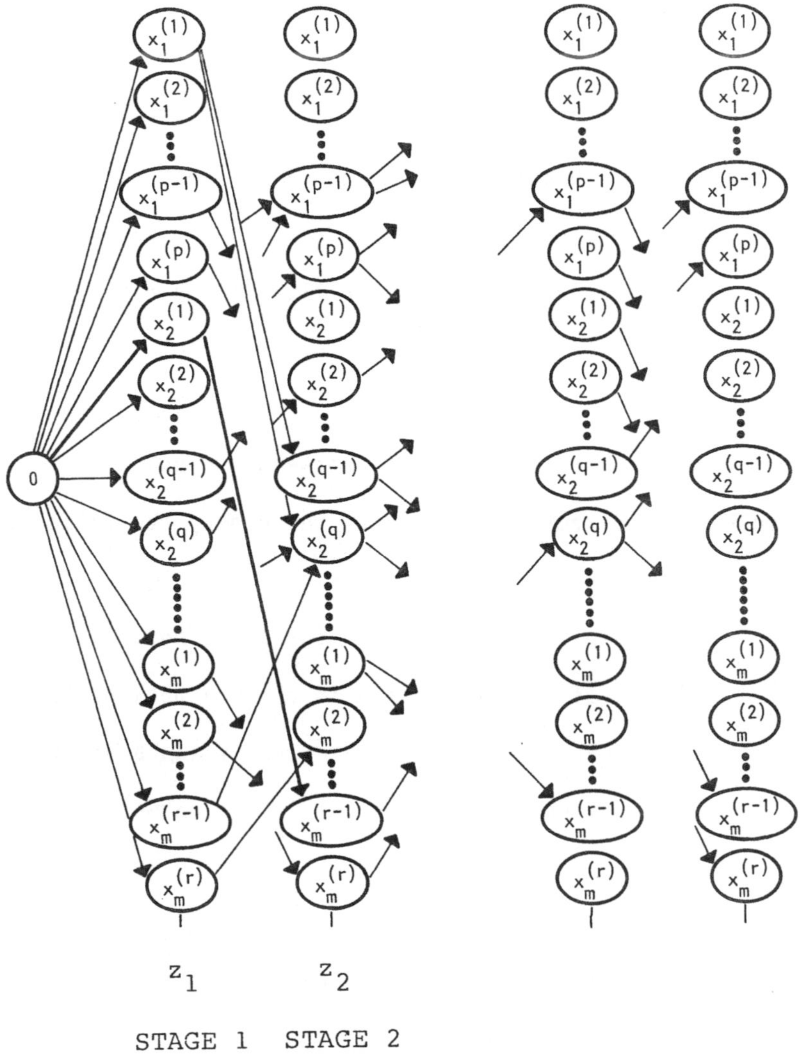

FIGURE 3.3 Network to solve the dispatching problem

The nodes on the network denote departures from specific airports at specific times. For example, node $X_2^{(1)}$ denotes a departure from airport X_2 at the first possible time (time $t_2^{(1)}$), while node $X_m^{(r-1)}$ denotes a departure from airport X_m at the $(r-1)$-th possible time (time $t_m^{(r-1)}$). The network only has oriented branches between individual nodes. The branch oriented from node $X_2^{(1)}$ towards node $X_m^{(r-1)}$ designates the possibility of the aircraft flying from city X_m at time $t_m^{(r-1)}$ upon completion of this flight to also fly from city X_m at time $t_m^{(r-1)}$. This means that in order for a branch between two nodes in the network to be directed from one to the other, the following conditions must be fulfilled :

(a) A branch is directed from the first node to the second if transportation is planned between the cities that represent the nodes (flights are only made between certain cities in the network) and if all planned flights have not yet been made from the first city.

(b) If the above condition is satisfied, the following equation must also be fulfilled :

$$t_i^{(p)} + t_{B_{i,j}} \leqslant t_j^{(q)} \qquad (3.4)$$

where :

$t_i^{(p)}$ - aircraft departure time from the i-th airport (in this case, the p-th possible)

$t_j^{(q)}$ - aircraft departure time from the j-th airport (the q-th possible)

$t_{B_{i,j}}$ - block flight time between airports i and j.

Figure 3.3. contains, in addition to other branches, a branch directed from node $X_2^{(1)}$ towards node $X_m^{(r-1)}$. This means that :

$$t_2^{(1)} + t_{B_{2,m}} \leqslant t_m^{(r-1)}$$

We denote respectively by $z_1, z_2, z_3, \ldots$ the first, second, third departure of the first aircraft. The first departure of the first aircraft can be one of departures $X_1^{(1)}, X_1^{(2)}, \ldots, X_1^{(p)}, X_2^{(1)}, X_2^{(2)}, \ldots, X_2^{(q)}, \ldots,$ $X_m^{(1)}, X_m^{(2)}, \ldots, X_m^{(r)}$. We note branch (z_{i-1}, z_i) that joins the $(i-1)$-th departure to the i-the departure. We denote the length of this branch by $R_i(z_{i-1}, z_i)$ which will stand for the profit on the i-t flight. This means that :

$$R_i(z_{i-1}, z_i) = c_i \cdot \lambda_i \int_{t-\tau}^{t+\tau} h_i(t)\,dt - C_i \qquad (3.5)$$

We denote by $f_i(z_i)$ the length of the longest path to node z_i that denotes the i-th departure of the first aircraft, i.e. the i-th stage. Since $i=0$ is the first stage, this is $f_0(0) = 0$. The dynamic programming equation reads :

$$f_0(0) = 0 \qquad (3.6)$$

$$f_i(z_i) = max\{f_{i-1}(z_{i-1}) + R_i(z_{i-1}, z_i)\} \qquad (3.7)$$

When $i=1$, with the assistance of equations (3.6) and (3.7) we calculate the optimal departure time that earns the maximum profit when only one flight is flown on the entire network.

In the next iteration for $i=2$ the departure times are calculated for two fights that earn the maximum profit when only two flights are flown on the entire network. Then we take the third iteration, etc. This example focussed on maximizing profit when determining departure times. It is clear that the airline schedule can also be designed with the goal of maximizing revenues, the total number of passengers flown or some other variable.

One aircraft can only fly a fixed number of flights. By routing the first aircraft, i.e. by determining the schedule of flights to be flown by the first aircraft, the following situation might appear. On some routes, not a single flight has been flown, on some routes some of the planned flights have been made, while on some routes all planned flights have been completed. Routes with all planned flights completed are excluded from further consideration. For the routes where not a single planned flight has been made, the functional time dependence of the passenger flow remains unchanged. This functional dependence changes for the case of a route that has already had some fights flown, since some of the passengers demanding transportation are already "served" by the flights that have been made. It is clear that those "served" passengers should be removed from the total number of passengers requesting transportation. Let us discuss the first case when the aircraft capacity surpasses the number of passengers on a route who want to travel at the time offered by the airline schedule of the first aircraft. In this case, all passengers wishing to travel during the attraction interval of the offered flight can be served so that the new passenger flow during the attraction interval equals zero (Figure 3.4).

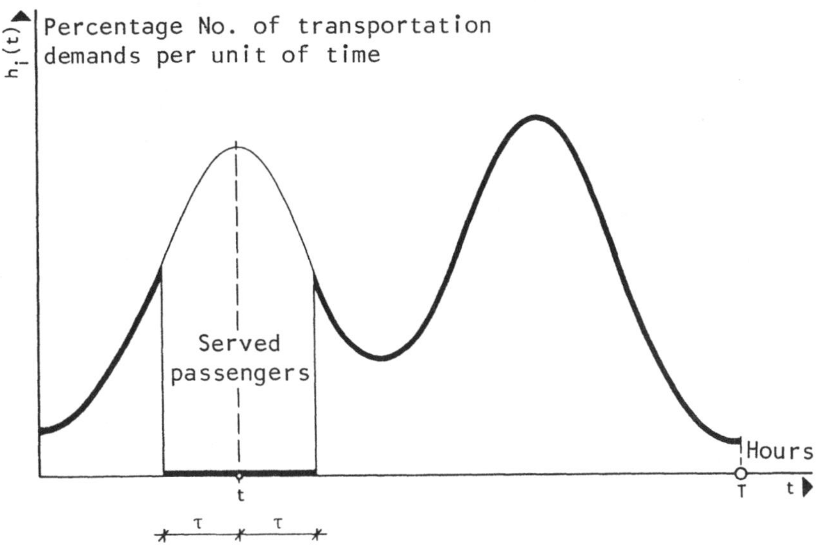

FIGURE 3.4. Functional time dependence of the passenger flow on route i after rotating the first aircraft (whose capacity is greater than the number of passengers wanting to travel)

When the aircraft capacity is smaller than the number of passengers wanting to travel on a route at the time offered by the first aircraft's airline schedule, some of these passengers will not be served. The passenger flow time dependence in this case is shown on Figure 3.5.

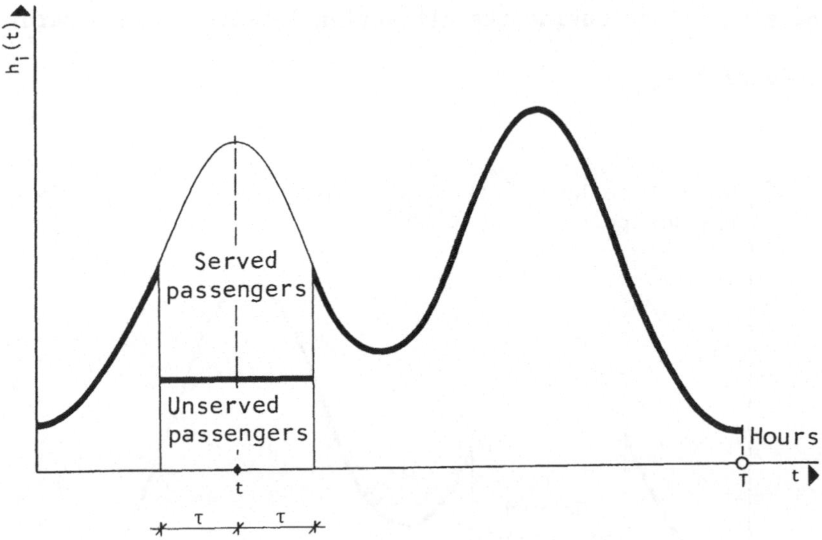

FIGURE 3.5 Passenger flow time dependence on route after routing the first aircraft (whose capacity is smaller than the number of passengers wanting to travel)

The procedure for designing the second aircraft's schedule is identical to the process for the first aircraft. When designing the schedule for the second aircraft, new passenger flow time dependencies are used. When the schedule has been completed for the second aircraft, the third aircraft is scheduled, etc. The airline schedule design is complete when all aircraft of a certain capacity have been assigned to routes and departure times have been calculated or when all available aircraft from the fleet have been assigned to routes (when aircraft of different capacities are used) and departure times are calculated, i.e. when all planned flights with a set number of aircraft have been completed.

The case we have just observed attempted to determine aircraft departures that would maximize profits. However, in certain cases the carrier must waive earning the maximum possible profit in order to satisfy operational constraints. For example, when designing the airline schedule, it might be required for the first aircraft to take off from the city in which the technical base is located, or to finish flying in a specific city. In the same vein, the city of origin or destination could be determined by some transportation policy, such as the carrier's desire to provide passengers good connections with the flights of other carriers, the desire for the new schedule to be similar to the schedule in the previous period, etc. The city of origin and departure times can also be stipulated by the meteorological conditions at certain airports.

When calculating flight block time between pairs of cities, the estimated time that the aircraft can spend waiting to land at the destination airport must be added to the total. Aircraft service time must also be calculated very carefully. Considerable traffic disturbances can arise if the service time anticipated in the airline schedule is less than what is realistically possible.

The described way to design an airline schedule is sequential in essence, since the schedule is designed in sequences - first for the first aircraft, then the second, etc. It is clear that the departure times obtained in this manner differ from those that would be obtained if the schedule was designed simultaneously for all aircraft. The sequential approach avoids considerable calculational difficulties that can arise in the simultaneous approach (particularly for larger transportation networks). However, we would repeat that solutions obtained in this manner are not optimal.

3.5. Determining the Least Number of Aircraft Needed on a Network With Known Departure Times

Aircraft departure times can be fixed in advance or takeoffs can take place within previously set time intervals. In the latter case, we distinguish the earliest, latest and desired departure times. As noted by Pollack [115] the desired departure time does not necessarily have to be in the middle of the earliest and latest departure times (Figure 3.6.b). Actual departures usually do not coincide with preferred times.

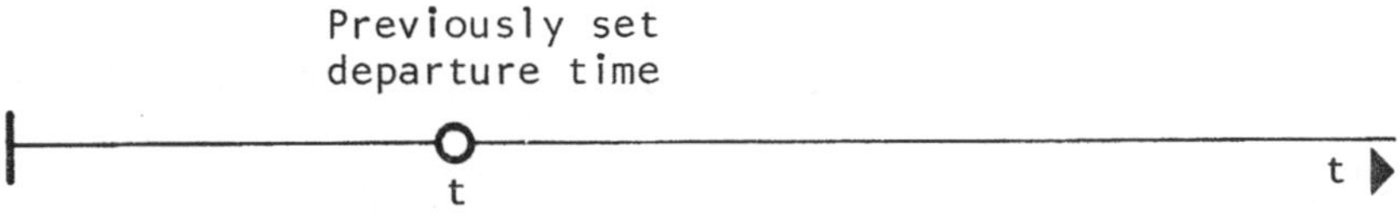

FIGURE 3.6 a   Previously set departure time

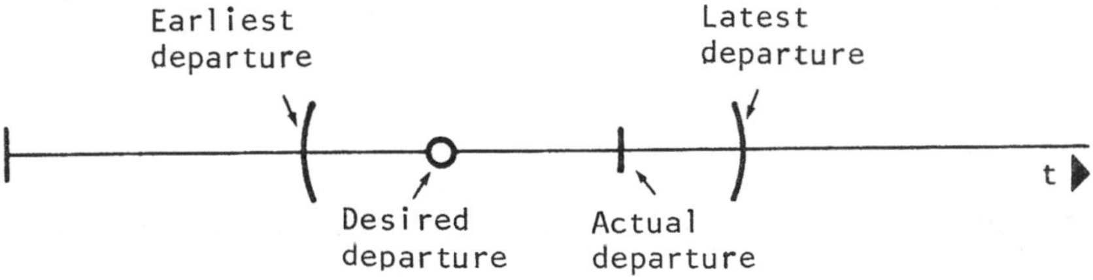

FIGURE 3.6 b   Previously set departure interval

Although departure time intervals enable greater flexibility in designing airline schedules, we will discuss the first case when departures on the network are fixed at advance (desired) times. The basic problem encountered in this case is determining the minimum number of aircraft needed to service the planned schedule. This problem is relatively easy to solve on smaller transportation networks. However, for larger networks the optimal solution must be chosen from the very large number of feasible solutions. A. Levin [90] presented a methodological procedure to determine the minimum number of aircraft needed to service a given schedule that can also be applied to other modes of transportation.

Before introducing Levin's method, let us discuss certain concepts needed to better understand the methodological process.*

An oriented network or oriented graph which does not contain a cycle is called an acyclic oriented graph.

Graph $G(N,A)$ whose set of nodes N can be divided into two subsets S and T so that

$$S \cup T = N \tag{3.8}$$

$$S \cap T = \emptyset \tag{3.9}$$

whereby the branches belonging to the set of branches A lead

---

* The following discussion is taken from the book : D. Teodorovic, "Transportation Networks", Gordon and Breach, London 1986.

from the nodes of set S towards the nodes of set T, is called a bipartite graph.

Bipartite graphs will be denoted as :

$$G = [S, P; A^*]$$

The term decomposing the acyclic oriented graph into chains means dividing the set of nodes N into subsets of nodes which do not have common components and which form chains with the corresponding branches that connect them. We would note that one isolated node can represent a chain as well, which means that acyclic oriented graph $G(N,A)$ can always be decomposed into $|N|$ number of chains, each one made up of only one node ($|N|$ is the number of nodes in set N). Figure 3.7 shows an acyclic oriented graph whose set of nodes contains nodes $x_1, x_2, \ldots, x_{20}$.

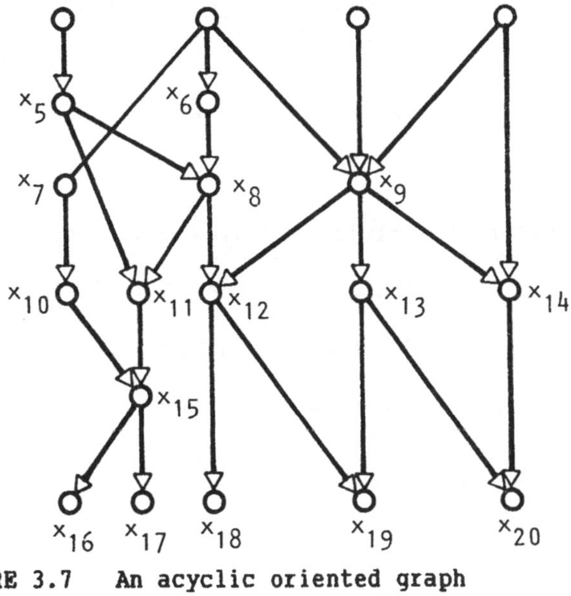

FIGURE 3.7 An acyclic oriented graph

The decomposition of this graph is presented on Figure 3.8. The graph is decomposed into 9 chains. Three chains are made up of only one node (chain $[x_6]$, chain $[x_{17}]$ and chain $[x_{19}]$).

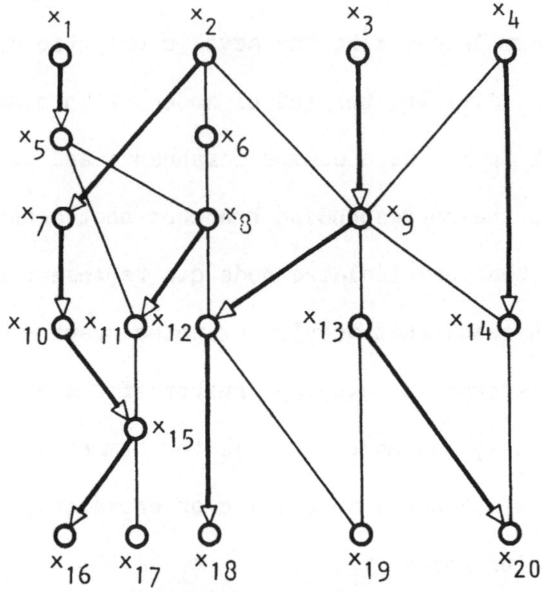

FIGURE 3.8 Decomposition of an acyclic oriented graph

An acyclic oriented graph can be decomposed into chains in several ways. It is clear that the larger the number of nodes included into individual chains, the smaller the number of chains into which the graph is decomposed. The following question arises when decomposing an acyclic oriented graph into chains : What is the smallest number of chains into which an acyclic oriented graph can be decomposed? However, before answering this question, let us return to our problem and show the connection between determining the minimum number of vehicles needed to service a given schedule on the transportation network and determining the minimum number of chains into which an acyclic oriented graph can be decomposed. We note Figure 3.9.

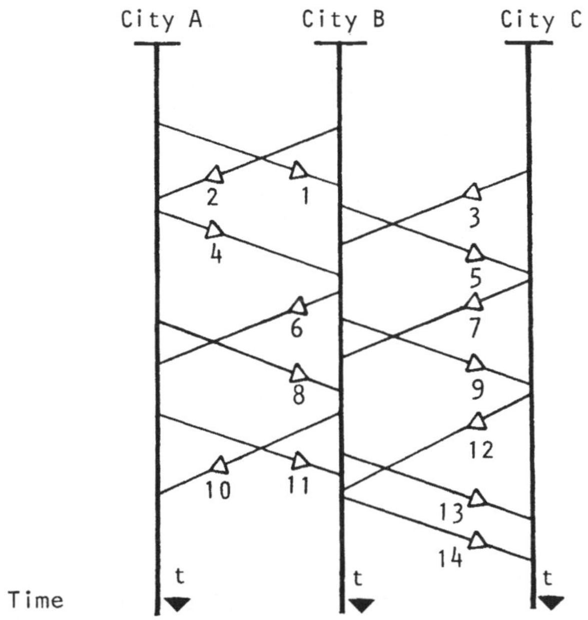

FIGURE 3.9 Space-time diagram

Figure 3.9 shows a space-time diagram with 14 flights to be carried out between cities A and B, cities B and C, cities C and B, and cities B and A. It is obvious that we can distribute the aircraft to service the 14 planned flights in different ways. For example, a vehicle can take flight 1, then flight 5, flight 7 and finally flight 10. If we understand flights to be nodes on the network, then chain (1,5,7,10) represents the route taken by the aircraft when carrying out planned flights 1,5,7 and 10. Figure 3.10 shows a network in which nodes represent planned flights.

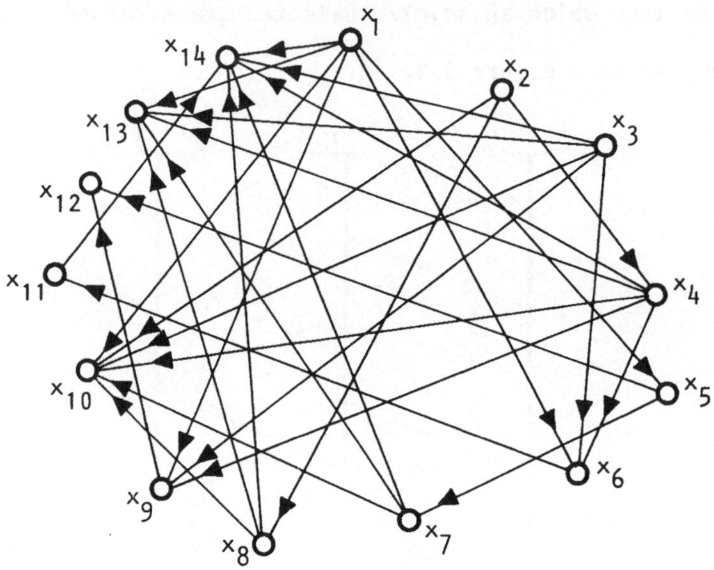

FIGURE 3.10 Network in which nodes represent flights

In this network, a branch is directed from node x_i towards node x_j only if flight x_j can be made after flight x_i. (Flight x_j can be made after flight x_i if flight x_j starts in the city where flight x_i finishes and if the planned departure time of flight x_j is after the arrival time of flight x_i).

Since chains represent vehicle routes, the minimum number of vehicles needed to service a given schedule on the transportation network equals the minimum number of chains into which the acyclic oriented graph can be decomposed, with each node representing flights to be made.

Therefore, by discovering the answer to our question about the minimum number of chains into which an acyclic oriented graph can be decomposed, we also answer the question on the least number of vehicles needed to service a given schedule.

Let us examine acyclic oriented graph $G(N,A)$. We designate the number of chains into which the graph is decomposed with $|C|$. The chains are denoted respectively by $k=1,2,\ldots,|C|$. The number of nodes belonging to the k -th chain is denoted by n_k. The total number of nodes in graph G is denoted by $|N|$. Since every node belongs to only one chain, we have :

$$n_1 + n_2 + \ldots + n_{|c|} = |N| \qquad (3.10)$$

and further :

$$|N| = \sum_{k=1}^{|c|} n_k = \sum_{k=1}^{|c|} n_k + (|C| - |C|) \qquad (3.11)$$

$$|N| = \sum_{k=1}^{|c|} (n_k - 1) + |C| \qquad (3.12)$$

The number of branches in every chain is 1 less than the number of nodes in the chain. Therefore, if n_k is the number of nodes in chain k , then n_k-1 is the number of branches in chain k. It is clear that :

$$\sum_{k=1}^{|C|} (n_k - 1) \qquad (3.13)$$

is the number of branches belonging to the chains into which graph G is decomposed.

We denote this number with $|D|$. This means :

$$|D| = \sum_{k=1}^{|C|} (n_k - 1) \tag{3.14}$$

or :

$$|N| = |D| + |C| \tag{3.15}$$

Since the number of nodes $|N|$ of graph G is fixed, we can minimize the number of chains $|C|$ into which graph is decomposed by maximizing the number of branches $|D|$ which belong to the chains into which graph G is decomposed.

We construct bipartite graph $G=(S,T;\ A^*)$ which corresponds to graph $G = (N,A)$ shown in Figure 3.11.

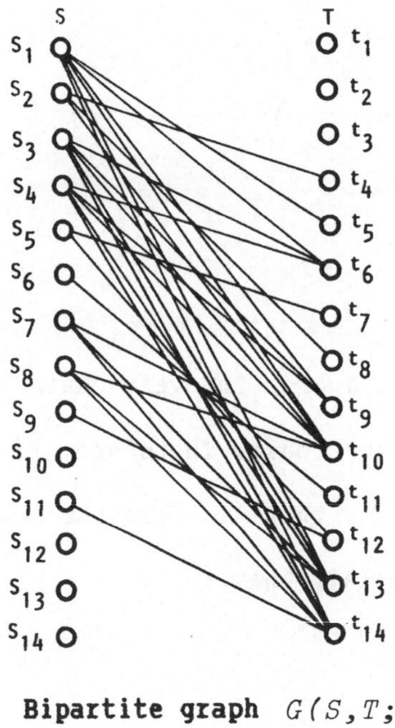

FIGURE 3.11 Bipartite graph $G(S,T;\ A^*)$

Since, for example, flight x_5 can be made after flight x_1, there is a branch in the corresponding bipartite graph $G(S,T; A^*)$ joining nodes s_1 with node t_5.

We assume that the capacity of every branch in the bipartite graph $(S_i, t_j) \in A^*$ equals 1. If branch (x_i, x_j) from initial graph G belongs to one of the chains into which graph G has been decomposed, then we note that a flow with a value of 1 goes through the corresponding branch on the bipartite graph. If branch (x_i, x_j) is not part of any of graph G's chains, then we note that there is no flow through the corresponding branch (s_i, t_j) on the bipartite graph, i.e. that the flow value equals 0. It is obvious that the opposite also holds true.

If the bipartite graph $G(S,T; A^*)$ contains a flow along branch (s_i, t_j), then branch (x_i, x_j) of graph $G=(N,A)$ is part of a chain into which graph G has been decomposed. This means that the total number of branches belonging to the chains of graph G, $|D|$, equals the total flow value going through the bipartite graph. Since some branches which are part of a chain belong to only that chain and some do not belong to any chain, it is clear that a flow with the greatest value of 1 can appear from all source nodes s_i, and that flows with the greatest value of 1 can arrive at all sink nodes t_j.

By minimizing $|C|$ or maximizing $|D|$ we maximize the total flows going through bipartite graph $G(S,T;\ A^*)$, keeping in mind that a flow with a maximum value of 1 can appear from every source s_i and a flow with a maximum value of 1 can arrive at every sink t_j.

<u>Example</u> : Maximize the flows in the example used to calculate the smallest number of vehicles needed to service the schedule shown in the space-time diagram in Figure 3.9. Figure 3.12 shows our bipartite graph one more time in which there are only branches starting from node s_1 .

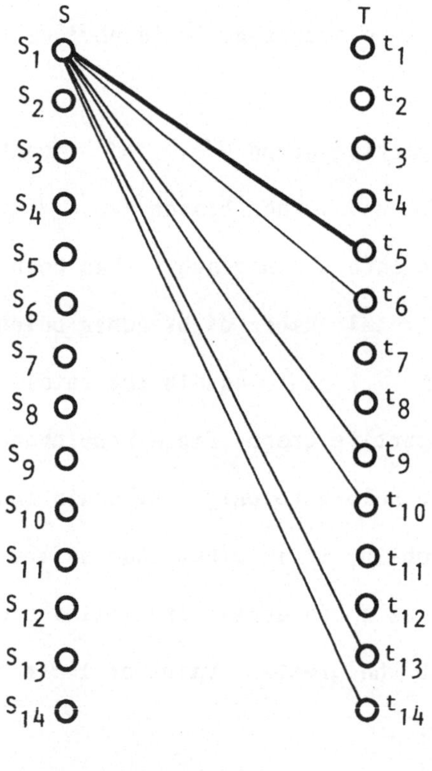

FIGURE 3.12 Bipartite graph

We maximize the flows through the bipartite graph in the following way. Begin with node s_1 and construct all branches starting from this node. These are branches (s_1,t_5), (s_1,t_6), (s_1,t_9), (s_1,t_{10}), (s_1,t_{13}), (s_1,t_{14}).

We allot a flow value of 1 to the first branch, branch(s_1,t_5). Since a flow with the highest value of 1 can appear from node s_1, this means that branches (s_1,t_6),(s_1,t_9),(s_1,t_{10}),(s_1,t_{13}) and(s_1,t_{14}) are left without a flow. The branch alloted the flow is denoted by a heavier line. We also note that in the future all branches arriving at node t_5 will be without a flow since node t_5 has already received a flow of 1.

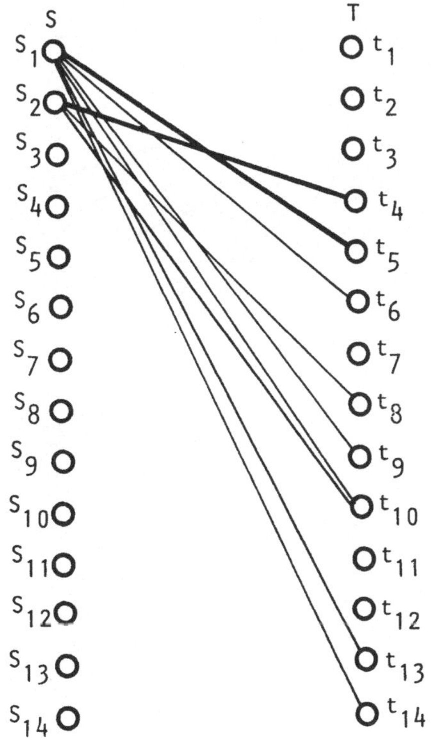

FIGURE 3.13 Allotting branch flows

Now we go to node s_2 . We construct branches (s_2,t_4) (s_2,t_8) and (s_2,t_{10}) from this node. The first branch, branch (s_2,t_4) is alloted a flow of 1. Branches (s_2,t_8) and (s_2,t_{10}) will be left without flows as will all future branches which enter node t_4 . (Figure 3.13).

The next node is s_3 . Branches starting from this node include branches (s_3,t_6), (s_3,t_9), (s_3,t_{10}), (s_3,t_{11}) and (s_3,t_{14}) . Branch (s_3,t_6) is alloted a flow with value of 1 and the other branches are left without flows (Figure 3.14).

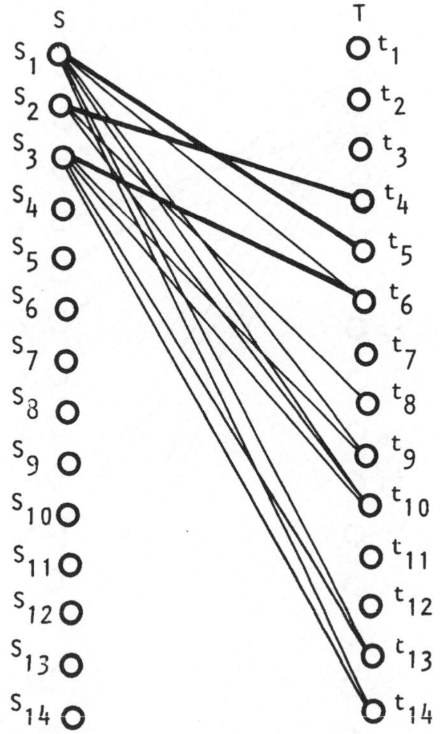

FIGURE 3.14 Allotting flows to branches

We now come to node s_4. Branches starting here include (s_4,t_6), (s_4,t_9), (s_4,t_{10}), (s_4,t_{13}) and (s_4,t_{14}). Up to now, as a rule, we have always allotted the first branch starting from the node with a flow value of 1. In this case, branch (s_4,t_6) cannot be allotted a flow since branch (s_3,t_6) arrives at node t_6 with a flow of 1. Therefore, we allot the flow to branch (s_4,t_9). Branches (s_4,t_{10}), (s_4,t_{13}) and (s_4,t_{14}) are left without flows (Figure 3.15).

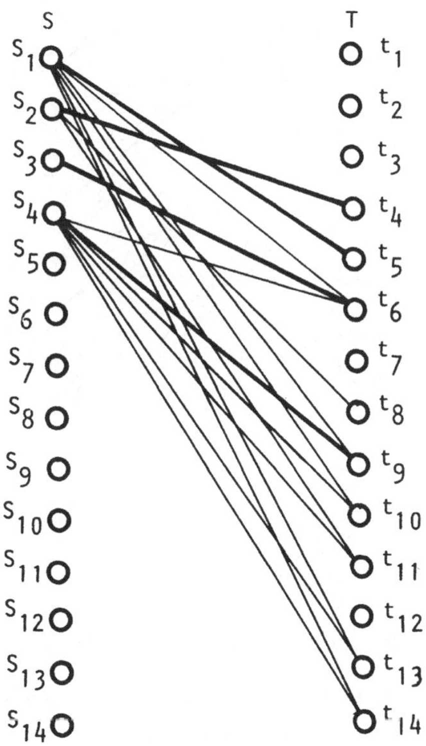

FIGURE 3.15 Allotting flows to branches

Continuing with our procedure, we allot flows of value 1 to branches (s_5,t_7), (s_6,t_{11}), (s_7,t_{10}),

(s_8,t_{13}), (s_9,t_{12}) and (s_{11},t_{14}).

The other branches in the bipartite graph will have intensity flows of zero. The final solution is shown on Figure 3.16.

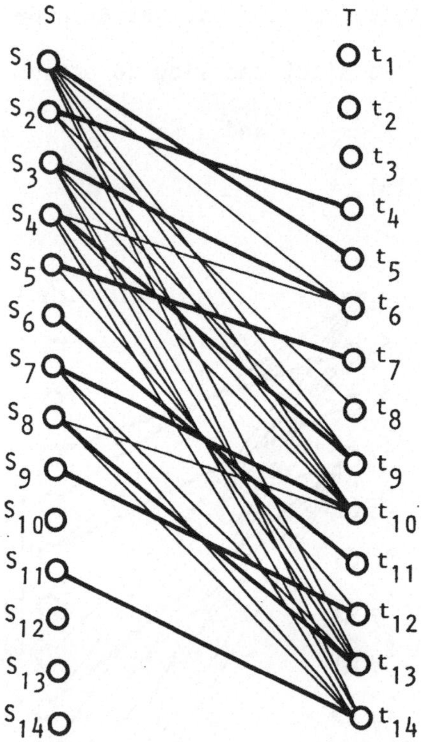

FIGURE 3.16 Optimal flows

As we can see from Figure 3.16, the bipartite graph has 10 branches with a flow value of 1. This means that the maximum flow through the bipartite graph is:

$$|D| = 10$$

Since the number of nodes in the initial graph was

$$|N| = 14$$

then the minimum number of vehicles needed to service the given schedule is :

$$|C| = |N| - |D| = 14 - 10 = 4$$

So, we can reliably say that the given schedule shown on the space-time diagram in Figure 3.16 can be serviced with 4 aircraft.

We must also design the routes for these aircraft. The bipartite graph on Figure 3.16 gives us the information concerning which branches become part of the chains into which the acyclic graph is decomposed. For example, the flow along branch (s_1, t_5) means that branch (x_1, x_5) in graph G becomes part of one of the chains into which the acyclic graph is decomposed. The acyclic graph's decomposition into chains, based on Figure 3.16 is shown on Figure 3.17.

Figure 3.18 shows the routes of the 4 vehicles.

The method described using a suitable computer program enables the construction of aircraft routing with the least possible number of aircraft. However, in spite of the fact that a computer program can be easily developed, some air carriers (mostly smaller companies) continue to construct weekly aircraft routing without the help of a computer. Compared to routing aircraft "by hand", the computer provides the following advantages in the airline schedule design phase :

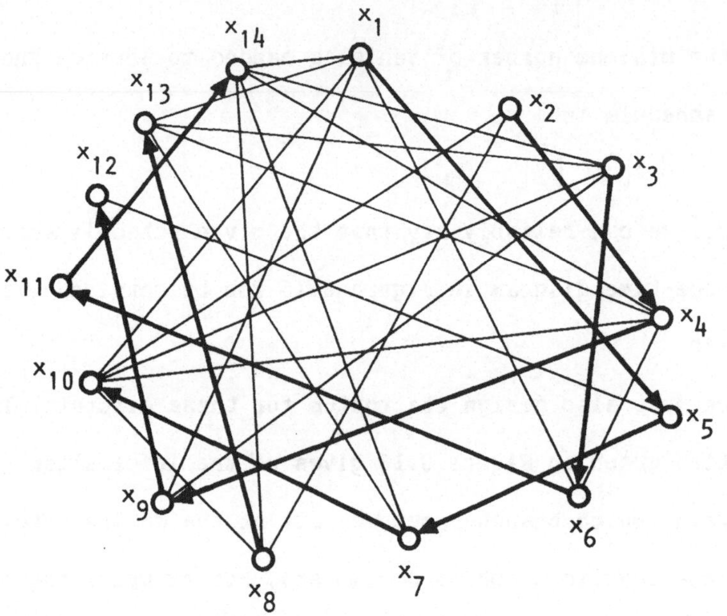

FIGURE 3.17 Decomposition of acyclic graph into chains

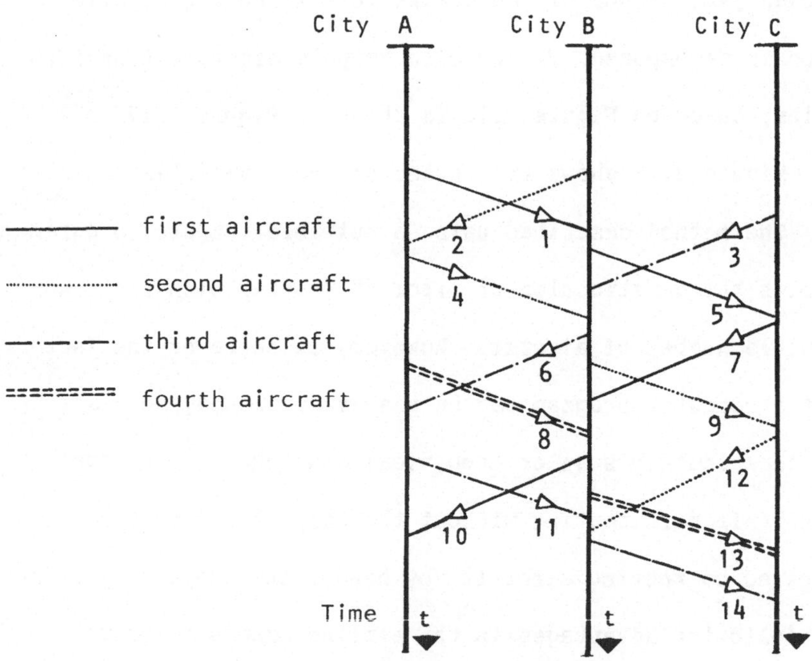

FIGURE 3.18 Routes of the four vehicles

1. Considerably decreases the time needed to elaborate the airline schedule and route aircraft,
2. Various airline schedules can be tested in a relatively short period of time, enabling the best one to be chosen,
3. The least number of aircraft needed for a given traffic volume can be reliably established,
4. As a function of the precisely formulated requirements of the other air carrier services, how the aircraft are routed can considerably facilitate the work of other departments, especially the flight crew and those in technical operations,
5. The margin for error in the airline schedule design and aircraft routing phase is considerably reduced,
6. Upon completion of a route, certain transportation indicators are immediately available, such as block flying time per aircraft, the weekly number of flights per aircraft, the greatest time interval between two flights, etc.

3.6 Method to Determine the Least Number of Aircraft Needed to Service a Network When There are Departure Time Windows

Our earlier discussions concerned aircraft rotation for previously established departure times so that planned flights are made with the least possible number of aircraft. Assigning fixed departure times on the routes considerably reduces the number of possible combinations to rationally rotate aircraft. The number of rotation combinations (from which the best should be chosen) rises substantially should planned departures on the routes be assigned within a specific time intervals. These time windows clearly must be based on market demand, or devised in consultation with commercial departments. The demands of technical operations and flight crews must also be studied along with other operational constraints when determining the time windows. The problems that arise with time windows include determining the least number of aircraft to service planned flights, determining departure times and assigning aircraft to specific routes. When solving these problems, we can strive to find either the optimal solution or a solution near to optimal. Due to the combinatorial nature of the problem and the fact that computer run time rapidly increases with the size of the problem, an optimal solution can be found primarily for smaller transportation networks. For larger networks, different heuristic procedures are used that are more efficient and give solutions near to optimal. Different operations research methods can be used to solve the problem of determining the minimum number of aircraft or designing aircraft routes when there are departure time intervals. These include combinatorial programming (branch-and-bound), dynamic or linear programming. This problem can also be treated as a set covering problem or as a problem to determine minimum total flow through certain branches of the network. The application of operations research methods, finding the optimal or near-optimal solution, depend on the specific transportation network, the carrier's policies, operational constraints, the size of the problem, etc. The works of Simpson,[146] Anders Martin-Lof,[94] Levin[90] and Pollack[115] are certainly among the most important that treat the problem of designing airline schedules and rotating aircraft when there are departure time intervals. This problem has incited considerable interest among transportation experts in recent years; it is found in all transportation branches and is solved using a similar methodological procedure. The most important works concerning schedule design when there are time intervals certainly include those of Desrosiers et al,[25][26][27][28] Psaraftis[122] and Jaw et a[168].

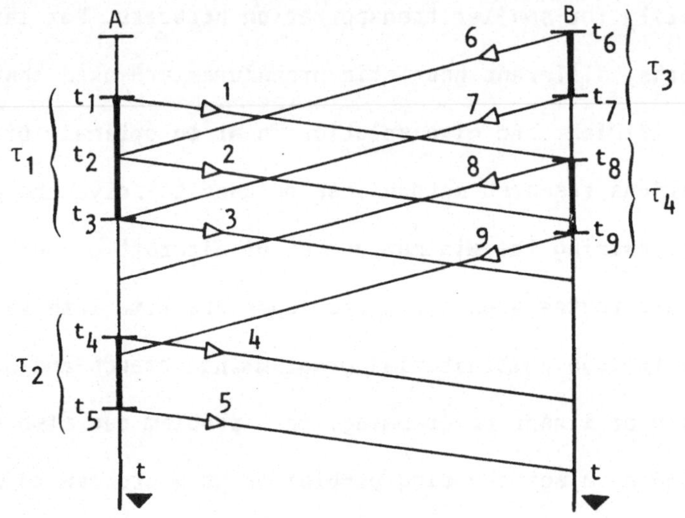

FIGURE 3.19 Space-time diagram when there are time intervals during which departures must take place

We will first illustrate the possibility of using integer linear programming to determine the minimum number of aircraft and design their routes when there are time intervals. Let us note Figure 3.19.

We assume that the first flight from city A to city B must start during time interval τ_1, the second flight from city A to city B during time interval τ_2, etc.

In order to reduce the size of the problem, we assume that the first departure from city A to city B can be made either at time t_1, at time t_2 or at time t_3. The second flight from city A to city B can be made at time t_4 or at time t_5, etc. Bipartite graph $G[S,T,A^*]$ corresponding to the space-time diagram on Figure. 3.19 is shown in Figure 3.20.

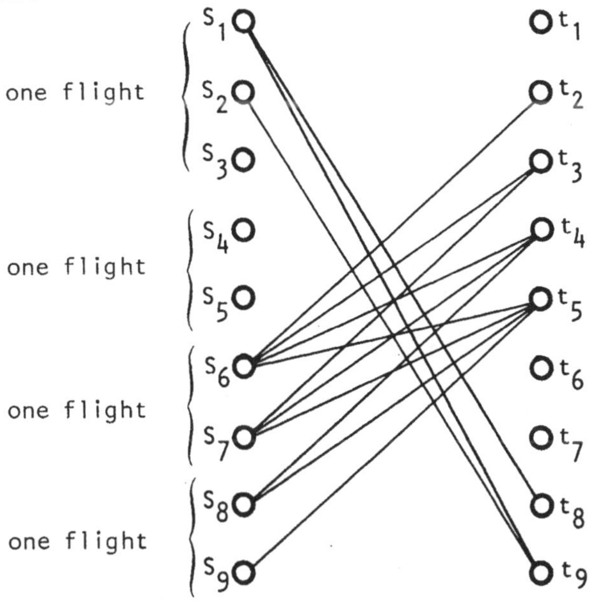

FIGURE 3.20 Bipartite graph $G|S,T,A^*|$

The first flight from city A to city B corresponds to nodes s_1, s_2, s_3, t_1, t_2 and t_3 in the bipartite graph. In this case, only one of nodes s_1, s_2, s_3 should be chosen. Should some node s_k be chosen, then its corresponding node t_k must also be chosen. Because of constraints of this type, the normal algorithm to find the maximum flow through the network cannot be applied. Therefore, we formulate our problem in another way as an integer linear programming problem.

We denote by N the set of flights to be flown. The number of flights to be flown $|N|$ should be different from the number of possible departures. For example, in the case of the space-time diagram shown on Figure 3.19, the number of flights to be flown $|N| = 4$, while there are nine possible departures (departures at times $t_1, t_2, t_3, \ldots, t_9$).

We introduce the following sets into the discussion :

$A(i) = \{j \mid (s_i, t_j) \in A^*\}$

$B(j) = \{i \mid (s_i, t_j) \in A^*\}$

$C_m = \{k \mid s_k$ belongs to the m-th flight

For the space-time diagram on Figure 3.19 and the bipartite graph on Figure 3.20, these sets contain the following components :

$A(1) = \{8,9\}$	$B(1) = \emptyset$
$A(2) = \{9\}$	$B(2) = \{6\}$
$A(3) = \emptyset$	$B(3) = \{6,7\}$
$A(4) = \emptyset$	$B(4) = \{6,7,8\}$
$A(5) = \emptyset$	$B(5) = \{6,7,8,9\}$
$A(6) = \{2,3,4,5\}$	$B(6) = \emptyset$
$A(7) = \{3,4,5\}$	$B(7) = \emptyset$
$A(8) = \{4,5\}$	$B(8) = \{1\}$
$A(9) = \{5\}$	$B(9) = \{1,2\}$

$C_1 = \{1,2,3\}$ $C_2 = \{4,5\}$ $C_3 = \{6,7\}$ $C_4 = \{8,9\}$

We denote the flow from node s_i to node t_j by x_{ij} , where :

$$x_{ij} = \begin{cases} 0, & \textit{if no flow from s to t} \\ 1, & \textit{if there is a flow from s to t} \end{cases}$$

Let us note the first flight from city A to city B. This flight can start either at time t_1, time t_2 or at time t_3 which means we will assign a flow to only one of the branches starting at nodes s_1, s_2 and s_3. We now introduce binary variable y_i defined as :

$$y_i = \begin{cases} 0, & \textit{if the i-th posible flight is not chosen} \\ 1, & \textit{if the i-th posible flight is chosen} \end{cases}$$

Since we will determine the least possible number of aircraft by maximizing the total flow from nodes s_i to nodes t_j, our objective function reads :

$$F = \sum_{m=1}^{|N|} \sum_{i \in C_m} \sum_{j \in A(i)} x_{ij} \rightarrow max \qquad (3.16)$$

Since only one flight departure time can be chosen out of all possible times, it is clear that :

$$\sum_{i \in C_m} y_i = 1, \quad \forall m \qquad (3.17)$$

We also have :

$$y_i \geqslant \sum_{j \in A(i)} x_{ij} \qquad (3.18)$$

$$y_i \geqslant \sum_{i \in B(j)} x_{ij} \qquad (3.19)$$

$$y_i, x_{ij} = 0,1, \forall\, i,j \qquad (3.20)$$

Levin[90] proposed the following procedure to solve this problem. If the constraint (3.20) is breached, and we no longer require for all variables y_i and x_{ij} to be either 0-1, this problem becomes one of linear programming. As mentioned by Levin, in a large number of cases linear programming produces an optimal solution that is an integer, so that the optimal solution obtained by applying linear programming is simultaneously the optimal solution to the original problem. However, should the linear programming not produce an integer, the optimal solution can be obtained by applying the branch-and-bound method.

Figure 3.21 shows the generation of a tree when using the branch-and-bound method (for more details of the method, see H. Taha[152]). The first node (node 1) shows values $y_1, y_2, \ldots, y_{|N|}$ obtained by applying linear programming where all variables y_i and x_{ij} are not required to be 0-1. We note the first variable that takes a value between 0 and 1. For node 1, let this be variable y_2, where $0 < y_2 < 1$. Let F_1 be the objective function value in node 1. We put out branches from node 1 to nodes 2 and 3. In node 2 variable y_2 takes value 0 and in node 3 value 1. We solve the linear programming problem once more for nodes 2 and 3. If the solution for node 2 or node 3 is an integer, the optimal solution is found. However, if there is no integer in either node 2 or node 3, further branching must be made. We denote respectively by F_2 and F_3 the objective function values in nodes 2 and 3. Let $F_2 < F_3$. Since we are striving to maximize the objective function, we continue branching from node 3. We once more find the next variable that takes a value between 0 and 1. Let this be variable y_k . This variable will take value 0 in node 4, and value 1 in node 5. We once more apply linear programming for nodes 4 and 5, etc.

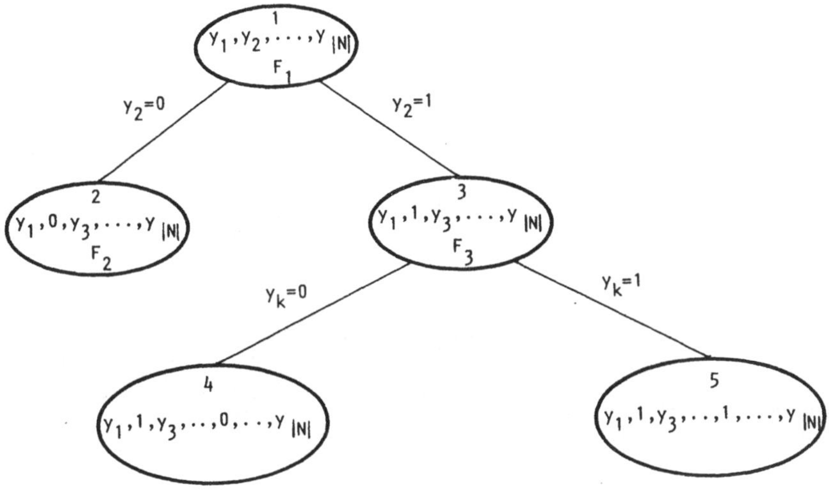

FIGURE 3.21 Generating a tree for branch-and-bound

As mentioned earlier, the problem of determining the minimum number of aircraft and their routes when there are time intervals during which departures must take place can be formulated as a set covering problem or as a problem of minimal total flow through branches of a network. Since it has been shown (Levin) that reducing this problem to one of a minimal total flow can considerably decrease the size of the problem, we will formulate our problem in such a way.

We denote by :

$A(i)$ - the set of nodes containing branches from node i ,

$B(i)$ - the set of nodes with branches going to node i ,

C_m - the set of nodes with branches that represent the possible departure of the m -th flight to be flown (Figure 3.22)

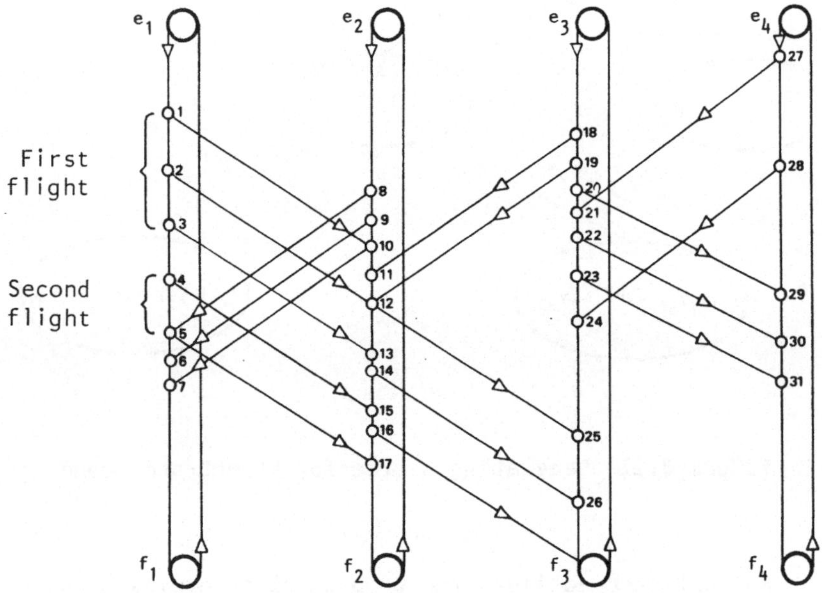

FIGURE 3.22 Minimizing total flow through network

Figure 3.22 presents the network where the minimal total flow is to be determined through branches denoting aircraft overnight stays at airports. In our example branches (f_1, e_1), (f_2, e_2), (f_3, e_3) and (f_4, e_4) denote the situation when aircraft spend the night at certain airports. The least number of aircraft spending the night at airports is actually the least number needed to service planned flights during the day.

We denote by f_{ij} the flow from node i to node j. If, for example, no aircraft is spending the night at the second airport, the flow intensity $f_{f_2 e_2}$ will equal zero. If an aircraft is spending the night at the second airport, then $f_{f_2 e_2} = 1$.

We ascertain that we will minimize the needed number of aircraft if we minimize the total flow through branches $f_{f_i e_i} (i = 1, 2, \ldots.)$, bearing in mind that for all flights to be flown, the flow can be "allotted" to only one branch that represents one of the possibilities for carrying out the flight. For example, in the case of the first flight (Figure 3.22), the flow will be allotted either to the branch starting from node 1, to the branch from node 2 or to the branch from node 3.

Our problem reads :

$$\sum_{(f_i, e_i)} f_{f_i, e_i} \rightarrow min \qquad (3.21)$$

$$\sum_{r \in B(i)} f_{ri} - \sum_{s \in A(i)} f_{is} = 0 \qquad (3.22)$$

$$\sum_{(i,j) \in C_m} f_{ij} = 1 \qquad (3.23)$$

where f_{ij} are integer variables.

In order to reduce the size of the problem, i.e. decrease the number of variables and number of constraints, Levin proposed making the following changes in possible departures and arrivals of certain flights before solving the problem :

1) Move all possible departures of some flight from any airport *A* upwards towards the nearest arrival time at that airport (Figures 3.23a and 3.23b)
2) Move all possible arrivals of some flight at any airport *B* down towards the nearest departure time from airport (Figures 3.24a and 3.24b).

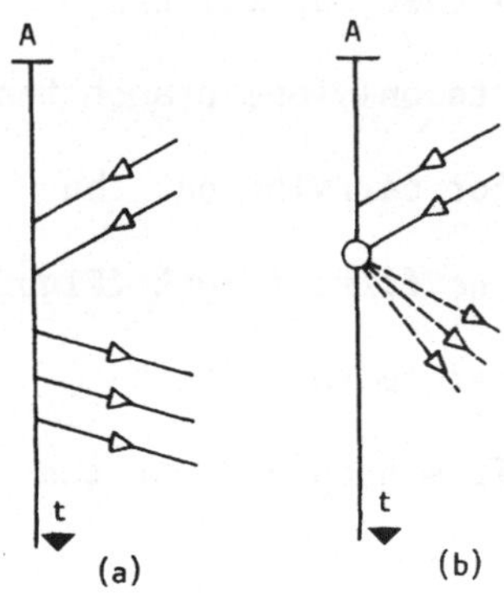

FIGURE 3.23 Moving departure towards closest arrival time

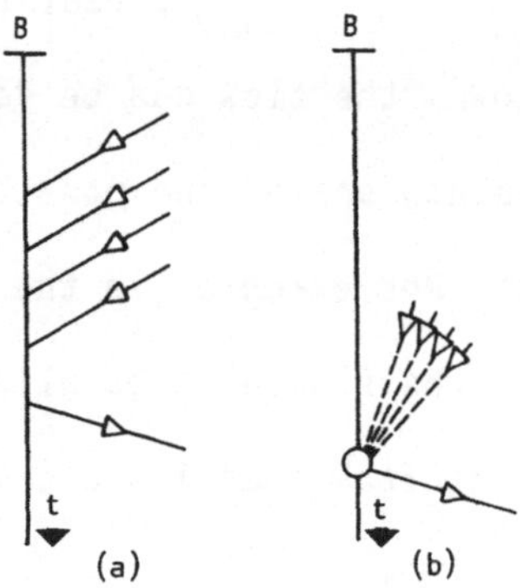

FIGURE 3.24 Moving arrival towards closest departure time

The first step is made by moving departure times towards the nearest arrival time, and then move the arrivals to the nearest departure. No changes are made in the nodes where the nearest departure and arrival are already "joined". (Figure 3.25).

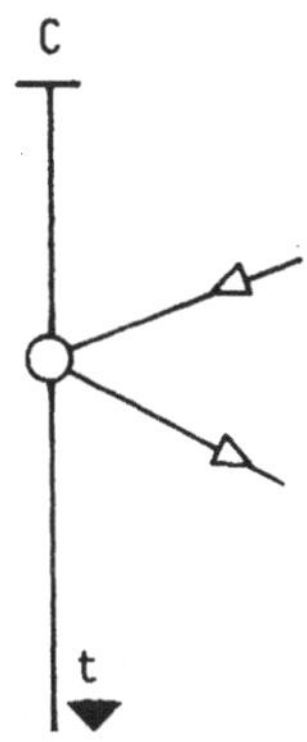

FIGURE 3.25 Nodes which are not moved

Since we are striving to minimize the number of aircraft needed to service planned flights, moving departures to the nearest arrival, or arrivals to the nearest departure is completely logical. We should consider how these movements affect a reduction in the size of the problem. By moving the arrival towards the departure (or vice versa), two nodes are joined together thereby reducing the number of constraints and variables by one. This considerably decreases the dimensions of the problem defined by equations (3.21), (3.22) and (3.23). We will illustrate the efficiency of this procedure in reducing the size of the problem of determining the least number of aircraft when departures must take place within preset time intervals.

The space-time diagram on Figure 3.26 shows flights that are to be flown during a specific time interval, whereby each flight corresponds to several moments in time during which the departure can take place.

After moving all possible departures upwards towards the nearest arrival time, the space-time diagram looks like Figure 3.27.

By moving all possible arrivals down towards the closest departure, the following space-time diagram is obtained (Figure 3.28).

As can be seen from Figures 3.26, 3.27 and 3.28, the procedure explained above considerably reduces the dimensions of the problem. For example, the total number of nodes on the network in Figure 3.26 is 122, while the total number of nodes on the network in Figure 3.28 is 36. This markedly decreases the number of constraints defined by equation (3.22). When the problem defined by equations (3.16) - (3.20) is solved by branch-and-bound, the routes of individual aircraft are determined along with determining the least number of aircraft. However, the size of the problem is rather large in this case. On the other hand, by defining the problem as one of minimizing the total flow, while using

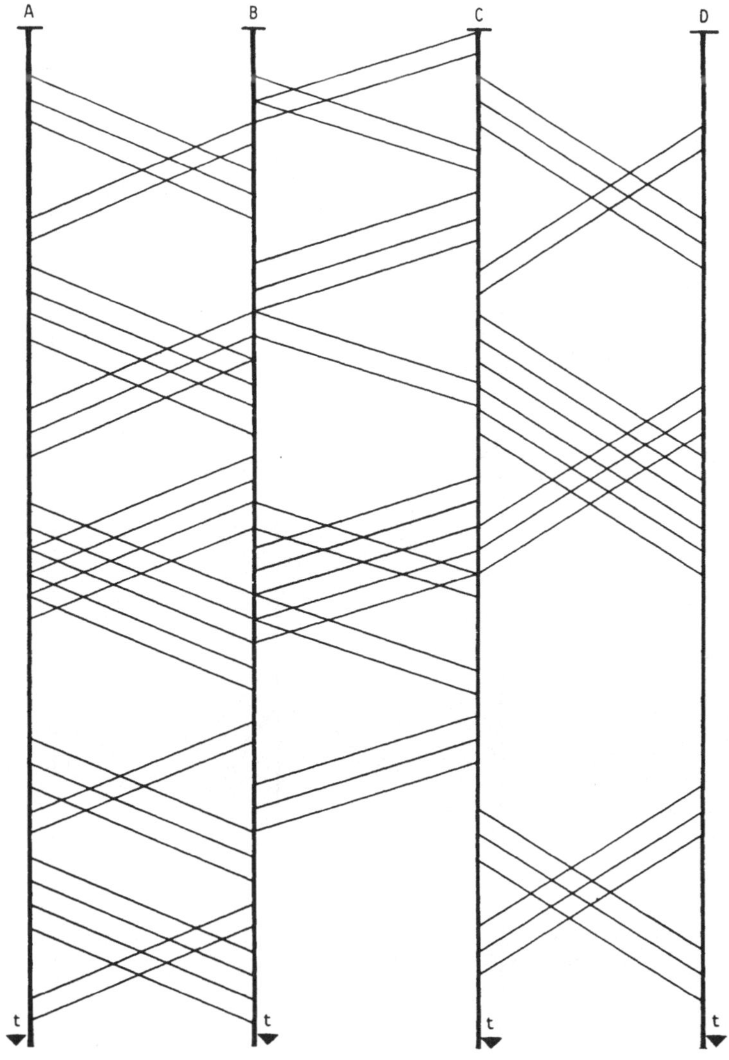

FIGURE 3.26 Space-time diagram where the number of nodes should be reduced

FIGURE 3.27 Space-time diagram after moving departures towards the nearest arrival times

FIGURE 3.28 Space-time diagram after moving arrivals towards the nearest departures

the procedure of moving departures to the nearest arrivals and vice versa considerably reduces the size of the problem. When treating the problem as one of minimal total flow, the routes of individual aircraft are not obtained immediately. These routes are relatively easy to determine based on the problem's solution.

The method described to find the least number of aircraft and design their routes when there are time intervals during which departures should take place becomes computationally inefficient for larger transportation networks. This inefficiency is both in terms of computer run time and in terms of needed capacity. As mentioned earlier, different heuristic procedures are used for larger networks which are computationally efficient and provide a solution that is near to optimal.

Pollack[115] showed that a relatively simple heuristic algorithm with a very short computer run time can be applied to determine the number of needed aircraft and their routes for a larger transportation network, i.e. a larger number of flights to be flown. It was previously mentioned that the departure time intervals should be determined by market demand or in consultation with commercial departments. Simpson[146] and Pollack[115] proposed the following procedure to determine departure time intervals. A primary pair of cities is determined for every flight with stopovers. These are actually the two cities between which the majority of the passengers fly. The functional dependence between passenger flow and time should be determined for every flight without stopovers and for every primary pair of cities with stopovers (Figure 3.29).

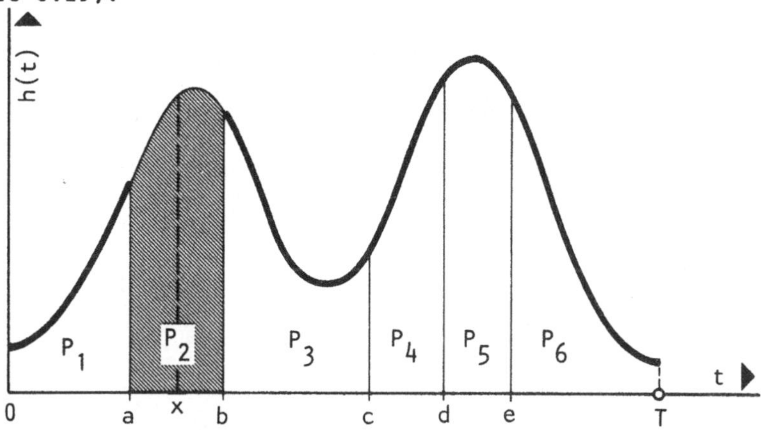

FIGURE 3.29 Functional dependence between passenger flow and time for a primary pair of cities

In earlier presentations it was mentioned that the shape of passenger flow is established on the basis of passenger surveys. The flight attraction boundary can be determined as follows. Let there be 6 flights to be flown between some primary pair of cities. The surface under curve $h(t)$ should now be divided into 6 approximately equal parts (Figure 3.29). Since:

$$P_1 = P_2 = P_3 = P_4 = P_5 = P_6 = \frac{1}{6} \int_0^T h(t)\,dt,$$

there will be approximately the same number of passengers on each flight. The flight attraction boundary is obtained by dividing the surface under curve $h(t)$ into the same number of parts as there is flight frequency. In this manner, every flight to be flown has a corresponding earlier and latest allowed departure time that are actually the interval boundaries within which departure should take place. Preferred departure times are determined within each interval as the arithmetic mean of desired departure times in the interval under observation (for example, time X in time interval (a,b). In other words, time X is determined from the equation :

$$X = \frac{a + b}{2} \tag{3.24}$$

The procedure described to determine time intervals and preferred times within them assumes that all flights are made with aircraft of the same capacity.. Should there be aircraft of different capacities, larger aircraft should take flights corresponding to greater passenger flows at the preferred departure time. It is clear that the different aircraft capacities make it necessary to redetermine the flight attraction boundary so that the surface under curve $h(t)$ corresponds to the capacity of the aircraft that will fly during that interval.

This procedure also simultaneously "distributes" passengers to individual flights. The passenger's choice of a particular flight is clearly a complex process and we can only roughly estimate the number of passengers on each flight in this manner. Each passenger chooses his flight independently of all other passengers. Many factors enter in to this decision, including the most important : duration of the flight or the type of flight (a flight with or without stopovers), departure times of competitive air carriers, lower air fares on certain flights (usually early morning or late night flights), schedules of alternative modes of transportation, preference for a certain type of aircraft, etc.

Should flights on a route take place only on certain days of the week, then these days must be determined in addition to the time intervals. This is best done in consultation with commercial services or with preliminary research into market demand. On some routes this is not an important point while on others the day of the flight can have a considerable influence on the number of passengers and the carrier's economic results. Therefore, determining the day on which flights are made in some situations requires lengthy study before an approach is made to designing the flight schedule.

The heuristic algorithm to determine aircraft departure times and calculate the least number of aircraft needed to service planned flights proposed by Pollack [115] is very simple. First, the preferred departure time is taken to be the departure time of all planned flights. Then the flights whose departures have been determined are linked into routes. Linking flights into routes is done using the FIFO rule (first in first out). Every flight coming into some city is linked to the first flight going out of that city (Figure 3.30).

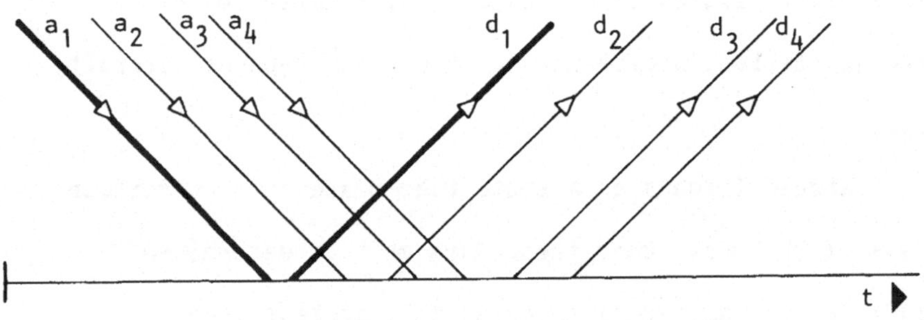

FIGURE 3.30 Linking incoming and outgoing flights

For example, the first incoming flight a_1 is linked to the first outgoing flight d_1 . Incoming and outgoing flights are linked for all cities in the network. If this had been previously set departure times instead of time intervals, the routes of each aircraft and the number needed to service planned flights would have been determined in this manner as well. However, since there are departure time intervals, it is clear that moving departure times within the intervals can reduce the needed number of aircraft. After the first phase when the decision is made to have the aircraft take off at preferred departure times and incoming and outgoing flights are linked for each city in the network, a certain number of flights are left unlinked (if no unlinked flights remain, then the needed number of aircraft and their routes have been found).

After the first phase, the number of unlinked incoming flights in each city equals the number of unlinked outgoing flights. In the next phase, an unlinked incoming and an unlinked outgoing flight are arbitrarily chosen and are linked to the effect that the arrival time of the incoming flight is moved up as much as possible, and the departure time of the outgoing flight is moved back as much as possible (Figure 3.31).

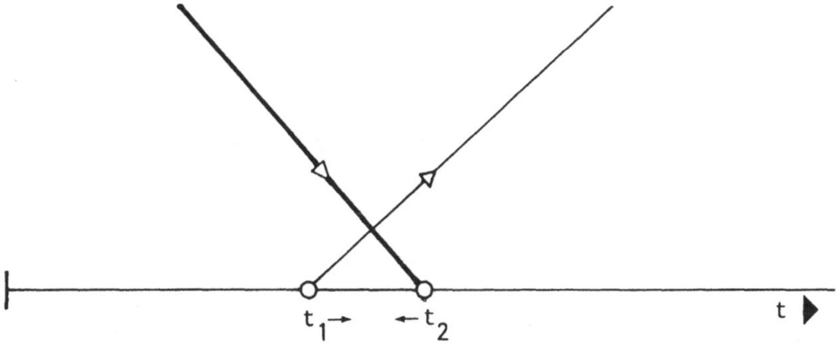

FIGURE 3.31 Linking an unlinked incoming flight and an unlinked outgoing flight

Unlinked flights are either completely isolated or else the first or last flight on the route being constructed.

If this concerns an isolated flight, linking is possible if, after being moved, the departure of the outgoing flight is after the arrival of the incoming flight and if the new departure times of the incoming and outgoing flights stay within the previously determined time intervals. If the unlinked flights are the beginning or end of a route, in addition to the above verification for isolated flights, another verification must be made so that moving the departure times does not interfere in the flights that have already been linked. It is clear that two unlinked flights can be linked only if departure times stay within the given time intervals and do not destroy already created routes. Should it prove impossible to link two arbitrarily chosen unlinked flights, two other flights are chosen. This procedure is completed when no new linkings are possible. This simultaneously determines departure times, the needed number of aircraft and their routes. When the procedure is completed, departures should be carefully studied along each route. In some cases, departures can be put as close as possible to preferred departure times without destroying already created routes.

The designed routes of the aircraft are normally shown in a space-time diagram or aircraft rotation plan. Figures

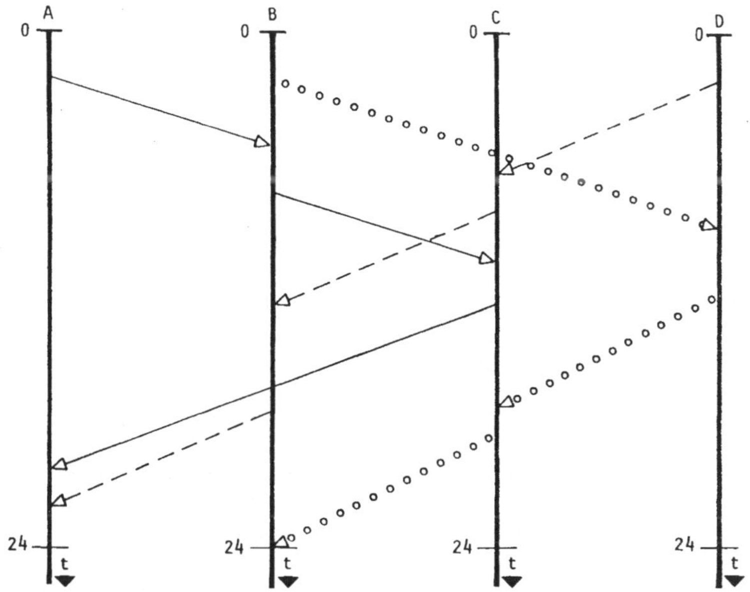

FIGURE 3.32 A typical space - time diagram (A,B,C,D - cities; t - time)

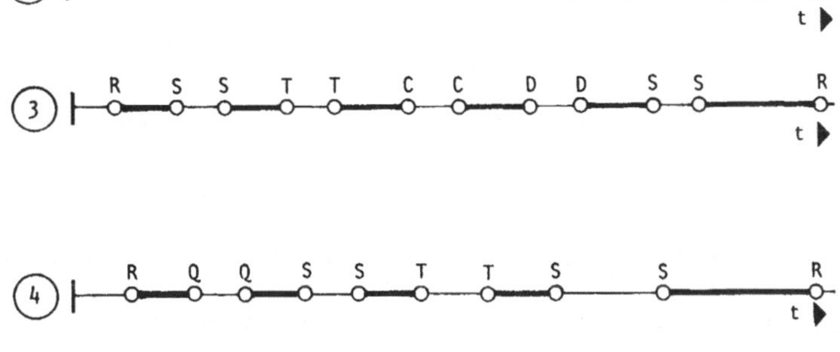

FIGURE 3.33 A typical rotation plan (1,2,3,4 - aircraft; t - time; A,B,C,... - cities)

3.32 and 3.33 present a typical space-time diagram and aircraft rotation plan.

The above procedure does not always provide the optimal solution and it is frequently shown that the planned traffic volume can be serviced with a smaller number of aircraft than that obtained by using the algorithm. The reason for this is primarily due to the heuristic nature of the algorithm. The goal of the procedure is not to establish complete automation in the phase of determining departure times, the needed number of aircraft and their routes. This algorithm, with suitable computer application, should only facilitate the airline schedule designer in his work. For example, by using this algorithm, minor changes to certain time intervals can be made that will reduce the number of aircraft needed to service planned flights.

3.7 The Effects of Meteorological Conditions on Airline Schedule Design

During the winter season, some airports are frequently closed to takeoff and landing operations due to meteorological reasons. For example, of the 1,753 cancelled JAT (Yugoslav Airlines) flights during 1983, 964 flights during the winter were cancelled due to meteorological reasons which is approximately 54% of all cancelled flights. At the same time, cancelled flights due to meteorological reasons during the summer season of 1983 made up only 15% of all cancelled flights. These facts alone indicate the need during the airline schedule design phase to pay considerable attention to meteorological conditions that govern the transportation network.

Networks are most often closed to take offs and landings during the winter because of lateral or tail winds that are above prescribed limits, poor visibility along the takeoff and landing runways and cloudiness. Only one or any combination of these factors can result in an airport being closed for a certain period of time.

In previous presentations we pointed out that air carriers usually try to adjust departure times to passenger demand as much as possible when designing airline schedules. However, if due consideration is not given to meteorological conditions at certain airports when designing the airline schedule, adjusting departures to passenger demand will not improve the quality of transportation services since many flights will be cancelled due to the weather. Clearly, the relationship must be established between meteorological conditions, air carrier economic results and the quality of transportation services in order to facilitate certain decisions regarding airline schedule design.

As already mentioned, meteorological conditions have a greater or lesser effect on the unimpeded operations of airports. The airport usability coefficient (the probability of an airport being open to take off and landing operations) is most often used as a global indicator of the level to which an airport is open to traffic. This probability is the quotient of the time during which an airport was open to take off and landing operations and the total time during the year during which the airport could have been open to traffic.

In order to understand the influence that meteorological conditions can have on designing an airline schedule, we first must establish how the probability of the airport being open to traffic varies during the months of the year and especially during the hour of day. In order to make this analysis, statistical data must be gathered over a period of 5-10 years. Should hourly variations in the probability that the airport will be open be insignificant, the airline schedule can be designed completely independent of meterological conditions. However, should these variations be more substantial, then meteorological conditions must clearly be taken into consideration when designing the airline schedule.

The influence of meteorological conditions on airport operations was discussed by Tosic and Babicl[66] among others. By studying the meteorological conditions at the Sarajevo Airport (Yugoslavia), Tosic and Babic established a function-

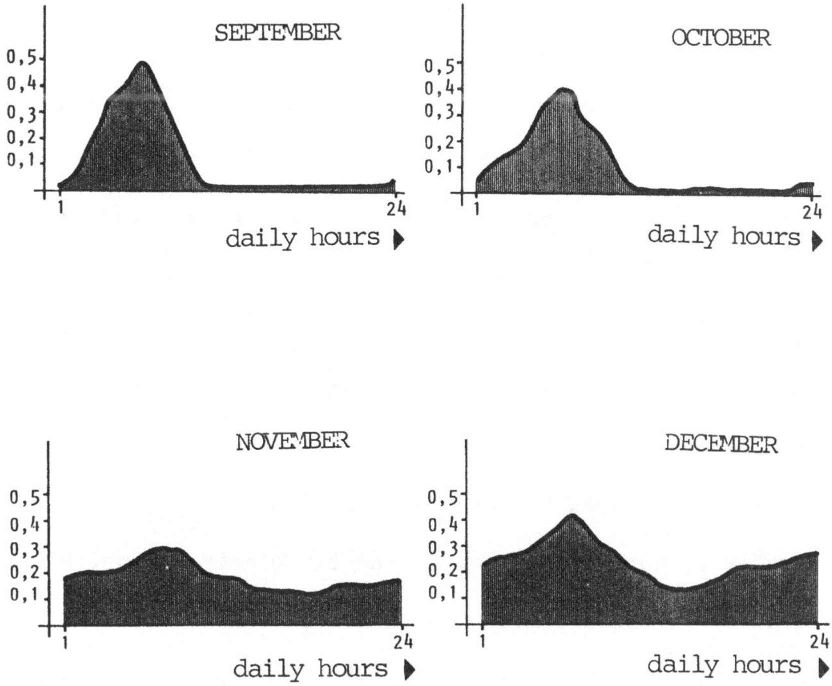

FIGURE 3.34 The dependence of the probability that the Sarajevo Airport will be closed to take off operations as a function of the time of day during the wintwr season

al dependence of the probability of the Sarajevo Airport being open to traffic based on the time of day. They made a separate calculation of the dependence between the probability that the airport will be closed to take offs and closed to landing operations as a function of the time of day. In order to calculate these probabilities at certain moments in time, the probability must be previously calculated that the airport will be closed due to tail winds, decreased visibility along the runway and/or low clouds. Tošić and Babić indicated, for example, that the probability of the Sarajevo Airport being closed to take offs or landings varied considerably by hour of the day during the winter months. Figure 3.34 presents the dependencies of the probability of the Sarajevo Airport being closed to take offs as a function of the time of day during the winter season.

Since the probability of airports on the network being closed to take off or landing operations can vary considerably during the day, making a poor choice of departure time can result in a large number of cancelled flights, thereby reducing the quality of the transportation services offered. We can conclude that meterological conditions governing the network are an essential constraint that must be taken into consideration when designing the airline schedule.

3.8 Designing an Airline Schedule With Maximum Reliability in Relation to Meteorological Conditions

In previous discussions of the problem of airline scheduling, we emphasized that passengers' flight demands must be given proper attention in order to adjust flight times to passengers' desires. We pointed out that every flight to be made normally corresponds to a specific time interval during which the flight should be made. Air carriers, from their point of view, are most often interested in flying their routes with the least possible number of aircraft, i.e. with the maximum possible utilization of their fleet. It is clear that several different aircraft routings can be made, each requiring an equal number of airplanes and each carrying out all flights within a given time interval. When discussing the effects of meteorological conditions on airline schedule designing, the problem of designing the airline schedule itself appears, i.e. producing an aircraft routing which allows for planned flights to be made with the least number of airplanes, and which has the greatest reliability regarding meteorological conditions of all aircraft routing plans which correspond to the least number of needed aircraft. This problem was discussed by Teodorović [160] who proposed the following method to solve it.

We assume that all transportation is made by airplanes with equal carrying capacity and that there is only one base in the network from which airplanes depart and to which they return. We are discussing the case of an airline schedule that repeats itself every day. During the day, let each airplane leave the base only once and then return to the base after completing its flight schedule. In other words, after returning to the base, not a single plane returns to the air.

We denote by P the set of flights to be flown, and by $|P|$ the total number of flights. Every flight y_i ($i=1,2,\ldots,|P|$) to be made is characterized by the city it takes off from, the city in which it lands, flight duration c_i, the expected number of passengers on the flight m_i and time interval $[a_i, b_i]$ during which the flight should depart. We understand flight duration in this case to be the sum of the block time (the time that elapses from motor start-up at the airport of origin to motor shutdown at the destination airport) and the time interval upon landing after which the airplane is ready for a new flight.

We now introduce graph $G(P,Q)$ into the discussion (Figure 3.35).

Branch (y_i, y_j) going from node y_i to node y_j belongs to the set of branches Q if flight y_j can be flown after completion of flight y_i. It is clear that flight y_j can be flown after flight y_i if flight y_i is completed before the

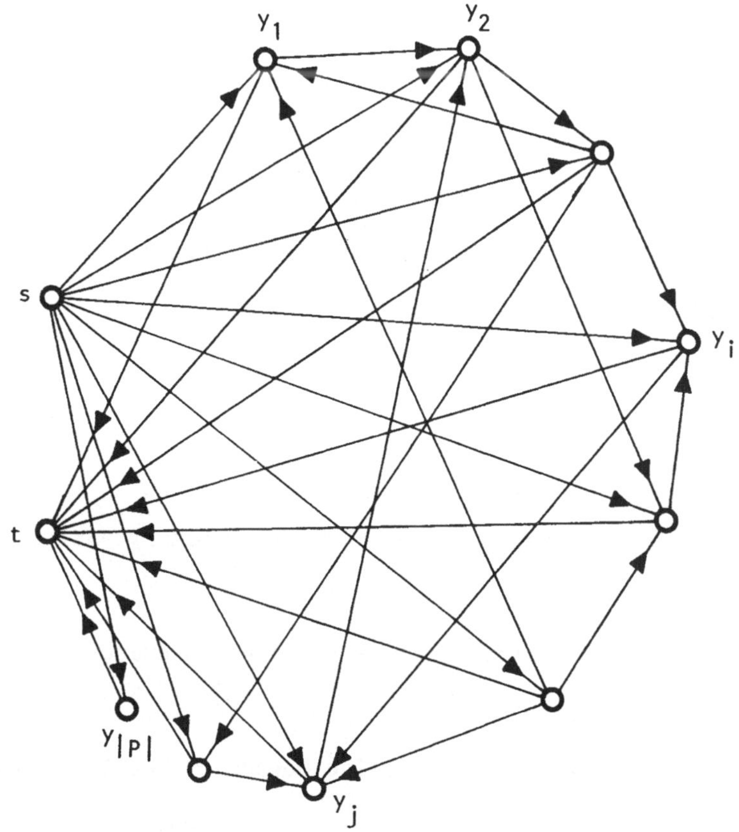

FIGURE 3.35 Graph G(P,Q) (The set of nodes comprising the flights to be flown)

latest possible take off time for flight y_j . We denote nodes that represent the exit and entrance to the base by s and t, respectively.

We assume that while carrying out planned flights "empty" flights (flights without passengers) are allowed and made. In other words, let there be branches directed from node s to all other nodes (the set of these branches is denoted by S) and branches directed from all nodes towards node t (the set of these branches is denoted by T). With C we denote the fixed costs for using an airplane. The network in which flights are made is defined by the set of nodes:

$$A = P \cup \{s\} \cup \{t\} \tag{3.25}$$

and the set of branches:

$$B = Q \cup S \cup T \tag{3.26}$$

We now introduce into the discussion binary variable defined as follows:

$$X_{ij} = \begin{cases} 1 \text{, when the airplane flies along branch } (y_i, y_j) \\ 0\text{, otherwise} \end{cases}$$

As already mentioned, airline companies are interested in carrying out their planned volume of flights with the least possible number of airplanes involved, so that the corresponding airline schedule has the greatest reliability in relation to meteorological conditions. Optimal routes for individual airplanes are obtained by solving the following problem:

$$C \sum_{y_j \in P} X_{sy_j} + \sum_{(y_i, y_j) \in Q} p_j m_j X_{ij} \to min \tag{3.27}$$

$$\sum_{y_j \in A} X_{ij} = \sum_{y_j \in A} X_{ji} = 1, \; y_j \in P \tag{3.28}$$

$$X_{ij} \geqslant 0, \; (y_i, y_j) \in B \tag{3.29}$$

$$t_i + c_i \leqslant t_j, \; (y_i, y_j) \in Q \tag{3.30}$$

$$a_i \leqslant t_i \leqslant b_i, \; y_i \in P \tag{3.31}$$

where p_j is the probability that flight y_j will be cancelled due to the weather.

So, we are striving to minimize the number of airplanes needed to execute planned flights. Should several solutions exist which correspond to the same number of airplanes, we are interested in the solution which corresponds to the smallest number of passengers whose flights could be cancelled due to the weather.

A similar problem was solved by Desrosiers et al.[25] They also minimized the total number of vehicles, and when several solutions existed that corresponded to the same number of vehicles, they chose the solution that had the smallest total travel time.

In our case, another term in the objective function is the expected total number of passengers whose flights might be cancelled due to the weather. Therefore, before solving this problem, we must first calculate probability p_j that flight y_j will be cancelled for meteorological reasons. This probability is:

$$p_j = 1 - \left[1 - p_{d_j}(t_j)\right]\left[1 - p_{a_j}(t_{f_j})\right] \quad (3.32)$$

where :

d_j - the airport where flight y_j begins,

a_j - the airport where flight y_j ends,

t_{f_j} - the time flight y_j ends,

$p_{d_j}(t_j)$ - the probability that the airport where flight y_j begins will be closed to traffic at time t_j ,

$p_{a_j}(t_{f_j})$ - the probability that the airport where flight ends will be closed to traffic at time t_{f_j} .

As can be seen, it is very easy to calculate probability p_j that flight y_j will be cancelled due to the weather assuming that the functional dependence of the probability of individual airports being closed during the day is known.

The expected total number of passengers whose flights could be cancelled due to the weather is :

$$\sum_{(y_i, y_j) \in Q} p_j m_j X_{ij} \quad (3.33)$$

Probability S that a passenger will not be able to make a planned flight due to the weather is:

$$S = \frac{\sum_{(y_j, y_j) \in Q} p_j m_j X_{ij}}{\sum_{y_j \in P} m_j} \quad (3.34)$$

Probability S is actually a percentage of the passengers whose flights could be cancelled due to the weather and is a very good indicator of the designed airline schedule's adaptability to existing meteorological conditions.

This problem is combinatorial by nature, and considerable computational difficulties can be expected when trying to find an optimal solution, especially for larger transportation networks or a larger number of flights to be made. For this reason, Teodorović [160] proposed the following heuristic algorithm for designing an airline schedule with the maximum reliability regarding meteorological conditions.

STEP 1 : Design all possible variations of the airline schedule that minimize total aircraft time on the ground, i.e. that minimize aircraft ground time between flights. Calculate the number of aircraft needed to execute these airline schedules.

STEP 2 : Calculate the total expected number of passengers whose flights could be cancelled due to the weather for all designed flight schedule variations and choose the variation having the least number of passengers.

The number of aircraft needed to carry out the airline schedule that minimizes the amount of aircraft ground time between flights can be determined using dynamic programming.

In our case, let individual flights be represented by stages (Figure 3.36).

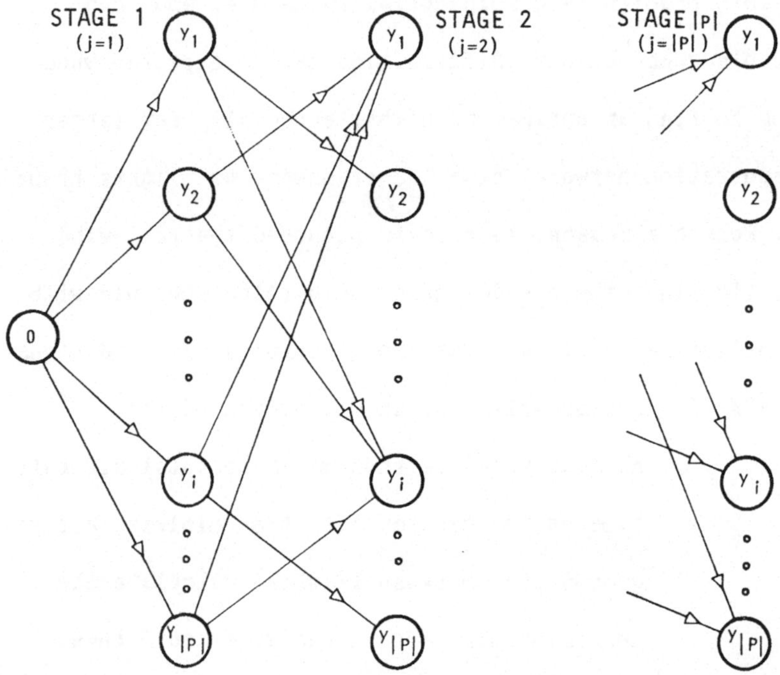

FIGURE 3.36 - Determining the number of airplanes needed using dynamic programming.

The first flight to be made, z_1 , represents stage 1, the second to be made, z_2 , is stage 2, etc. The first flight to be made, z_1 , could be flight y_1 or flight y_2 ..., or flight $y_{|P|}$. Graph $G(P,Q)$ gives us information as to which flights can be made after a specific flight has been made.

Actually, this graph only gives us the initial information. When constructing individual branches in the network as shown in Figure 3.36, we must bear in mind not only the possibilities of a particular flight being able to be flown after a specific flight, but also other operational restrictions connected to daily aircraft utilization, the working hours of different airports in the network, and similar concerns.

Let us take some branch (z_{i-1}, z_i). We denote its length by $R_i(z_{i-1}, z_i)$. Let this branch length represent the least amount of time that an aircraft must spend on the ground after completing flight z_{i-1} in order to be ready for flight z_i . We denote by $f_j(z_j)$ the length of the shortest path to node z_j located in stage j . Since the beginning stage is $j=0$, then $f_0(0)=0$. The dynamic programming relation reads :

$$f_0(0) = 0 \tag{3.35}$$

$$f_j(z_j) = \min\{f_{j-1}(z_{j-1}) + R_j(z_{j-1}, z_j)\} \tag{3.36}$$

$$j = 1, 2, \ldots, |P|$$

Based on relations (3.35) and (3.36), it is possible to determine different airline schedules which minimize the total time aircraft spend on the ground between individual flights.

Let us now discuss a numerical example of designing an airline schedule with maximum reliability regarding meteorological conditions. (The example is taken from D. Teodorovic). We note the transport network shown on Figure 3.37. There are 7 airports in the network. Let the air carrier's base be located in city A_1 from which all aircraft must take off and return to after completing their flights. Thirty flights are planned, denoted respectively by $y_1, y_2, \ldots, y_{30}$. Each aircraft leaves the base only once a day and all aircraft must return to the base by 23:00.

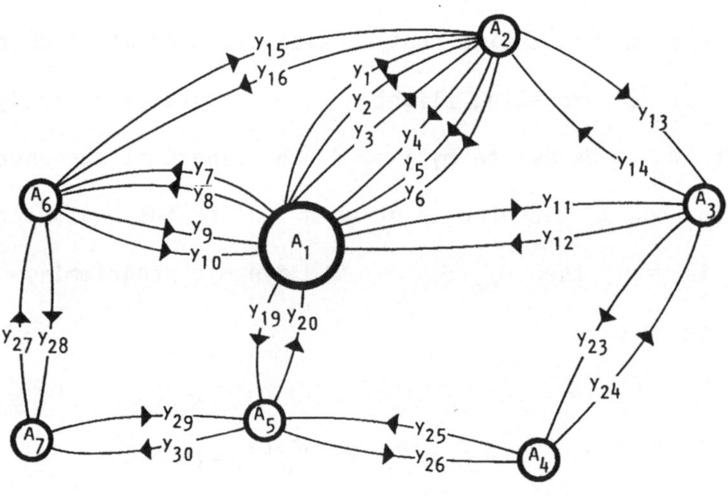

FIGURE 3.37 Network to maximize reliability regarding meterological conditions

We assume that "empty" flights, i.e. flights without passengers, are allowed when an airplane leaves or returns to the base. Flight times (in hours) between pairs of cities are given in the following matrix:

	A_1	A_2	A_3	A_4	A_5	A_6	A_7
A_1	0	1,5	1	1	1,8	1,1	2
A_2	1,5	0	1,4	2,4	2,7	1,7	2,8
A_3	1	1,4	0	1,2	2,6	3	2,9
A_4	1	2,4	1,2	0	1,1	2,8	2,5
A_5	1,8	2,7	2,6	1,1	0	1	1,5
A_6	1,1	1,7	3	2,8	1	0	1
A_7	2	2,8	2,9	2,5	1,5	1	0

The functional dependencies of the probability that airports will be closed to take off or landing traffic during the day are given in Table 3.1.

Airport	*Functional dependence of the probability that it is closed to takeoffs*	*Functional dependence of the probability that it is closed to landings*
A_1	$P_1 = \begin{cases} -0.0083t & 0<t\leq 6, \\ -0.025t+0.2 & 6<t\leq 8 \\ 0 & 6<t\leq 24 \end{cases}$	$S_1 = \begin{cases} 0.0166t & 0<t\leq 6, \\ -0.0166t+0.2 & 6<t12 \\ 0 & 12<t\leq 24. \end{cases}$
A_2	$P_2 = \begin{cases} 0.0333 & 0<t\leq 6, \\ -0.0333t+0.4 & 6<t\leq 12, 12<t\leq 24 \\ 0 & \end{cases}$	$S_2 = \begin{cases} 0.0142t & 0<t\leq 7 \\ -0.025t+0.275 & 7<t\leq 11, \\ 0 & 11<t\leq 24. \end{cases}$
A_3	$P_3 = \begin{cases} 0.08t & 0<t\leq 8, \\ -0.083t+1.27 & 8<t\leq 14 \\ -0.01t+0.24 & 14<z\leq 24 \end{cases}$	$S_3 = \begin{cases} 0.071t & 0<t\leq 7, \\ -0.083t+1.081 & 7<t\leq 13, \\ 0 & 13<t\leq 24 \end{cases}$
A_4	$P_4 = \begin{cases} 0.022t+0.2 & 0<t\leq 9 \\ -0.1t+1.3 & 9<t\leq 11 \\ 0.2 & 11<t\leq 24. \end{cases}$	$S_4 = \begin{cases} 0.1 & 0<t\leq 8, \\ 0.1t-0.7 & 8<t\leq 10, \\ -0.014t+0.443 & 10<t\leq 24. \end{cases}$
A_5	$P_5 = \begin{cases} 0.011t+0.1 & 0<t\leq 9, \\ -0.0067t+0.26 & 9<t\leq 24. \end{cases}$	$S_5 = \begin{cases} 0.11t+0.1 & 0<t\leq 9, \\ -0.0067t+0.26 & 9<t\leq 24. \end{cases}$
A_6	$P_6 = \begin{cases} 0.0375t, & 0<t\leq 8, \\ -0.019t+0.45 & 8<t\leq 24, \end{cases}$	$S_6 = \begin{cases} 0.025t & 0<t\leq 8, \\ -0.02t+0.36 & 8<t\leq 18, \\ 0 & 18<t\leq 24. \end{cases}$
A_7	$P_7 = \begin{matrix} 0.0444t & 0<t\leq 9, \\ -0.027t+0.64 & 9<t\leq 24. \end{matrix}$	$S_7 = \begin{matrix} 0.0444t & 0<t\leq 9, \\ -0.027t+0.64 & 9<t\leq 24. \end{matrix}$

Flight	*Average No. of passengers*	*Time interval in which flight must take off*
y_1	100	6 - 8
y_2	80	8 - 18
y_3	100	18 - 22
y_4	100	6 - 8
y_5	80	8 - 17
y_6	100	17 - 22
y_7	80	6 - 10
y_8	70	10 - 22
y_9	80	6 - 10
y_{10}	70	10 - 22
y_{11}	60	6 - 22
y_{12}	60	6 - 10
y_{13}	110	6 - 22
y_{14}	110	6 - 22
y_{15}	60	6 - 22
y_{16}	60	6 - 22
y_{17}	100	6 - 22
y_{18}	100	6 - 22
y_{19}	110	6 - 22
y_{20}	110	6 - 22
y_{21}	60	6 - 22
y_{22}	60	6 - 22
y_{23}	90	6 - 22
y_{24}	90	6 - 22
y_{25}	70	6 - 22
y_{26}	70	6 - 22
y_{27}	60	6 - 22
y_{28}	60	6 - 22
y_{29}	110	6 - 22
y_{30}	110	6 - 22

Table 3.2. gives the expected number of passengers per flight and the intervals during which take off should occur.

In order to solve this problem, Teodorovic developed a computer program. The first part of the program generates airline schedules with the minimum total aircraft ground time between two consecutive flights. The second part of the program chooses the schedule with the least number of passengers whose flights would be cancelled due to meteorological conditions out of all possible airline schedules requiring an equal (smallest possible) number of aircraft.

For the numerical example under consideration, 4 different aircraft routing schedules were obtained, i.e. 4 airline schedules that could be made using 4 aircraft (the smallest number of aircraft possible). These aircraft routings are shown in Table 3.3.

First possible aircraft routing	
Aircraft	*Flights flown*
I	$Y_{27}, Y_9, Y_2, Y_5, Y_8, Y_{10}, Y_{11}, Y_{12}, Y_{19}, Y_{18}, Y_{17}, Y_{20}, Y_{21}$
II	$Y_1, Y_4, Y_7, Y_{15}, Y_{13}, Y_{14}, Y_{16}, Y_{28}, Y_{29}, Y_{26}, Y_{24}$
III	Y_{22}, Y_3, Y_6
IV	Y_{23}, Y_{25}, Y_{30}

Second possible aircraft routing	
Aircraft	*Flights flown*
I	$Y_{22}, Y_1, Y_5, Y_2, Y_{13}, Y_{12}, Y_8, Y_{10}, Y_{11}, Y_{14}, Y_6, Y_3$
II	$Y_4, Y_7, Y_9, Y_{19}, Y_{18}, Y_{17}, Y_{20}, Y_{21}, Y_{24}, Y_{23}, Y_{25}, Y_{26}$
III	$Y_{15}, Y_{16}, Y_{28}, Y_{27}$
IV	Y_{29}, Y_{30}

Third possible aircraft routing	
Aircraft	Flights flown
I	$Y_{29}, Y_{18}, Y_{17}, Y_{20}, Y_8, Y_{10}, Y_{11}, Y_{12}, Y_2, Y_5, Y_3, Y_6$
II	$Y_1, Y_4, Y_7, Y_{15}, Y_{13}, Y_{14}, Y_{16}, Y_{28}, Y_{27}$
III	$Y_9, Y_{19}, Y_{26}, Y_{24}, Y_{23}, Y_{22}, Y_{21}$
IV	Y_{25}, Y_{30}

Fourth possible aircraft routing	
Aircraft	Flights flown
I	$Y_{25}, Y_{18}, Y_{17}, Y_{20}, Y_8, Y_{10}, Y_{11}, Y_{12}, Y_2, Y_5, Y_3, Y_6$
II	$Y_1, Y_4, Y_7, Y_{15}, Y_{13}, Y_{14}, Y_{16}, Y_{28}, Y_{27}$
III	$Y_9, Y_{19}, Y_{26}, Y_{24}, Y_{23}, Y_{22}, Y_{21}$
IV	Y_{29}, Y_{30}

The smallest probability S corresponds to the first aircraft routing (S = 0.24).

In the above numerical example, the airline company had only one base. It was also assumed that "empty " flights were allowed and that all aircraft had the same carrying capacity. This model can be modified for networks with several bases and for aircraft with different carrying capacities. As already mentioned, probability S representing the percentage of passengers whose flights would be cancelled due to weather conditions is a good indicator of the designed airline schedule's adaptability to meteorological conditions. This probability can be decreased by increasing the number of aircraft in use, by changing the time intervals during which take off must take place, by opening new bases in the network, etc. The functional dependence of the probability of individual airports being open for take off and landing traffic during the day is the basic input data needed to apply the above model. In order to establish these dependencies, a statistical data base is needed covering the previous 5-10 years.

3.9. The Influence of the Aircraft Maintenance System on Designing an Airline Schedule

In previous discussions of the airline schedule problem, it was stressed that one of the most important operational limitations that can have a crucial effect on the choice of alternative airline schedules is the system of aircraft maintenance. It was also emphasized that due to flight security reasons, the inspection or replacement of certain parts must be done at a certain time and at a certain airport that is most often the air carrier's technical base. The fact that aircraft have to be removed from operations for a lengthy period of time due to larger technical work could have an important influence on the choice of airline schedule variation.

With a view to including maintenance department (and other essential department) requirements into the airline schedule, most international air carriers do the following. The airline schedule design department sends a first variation (or several variations) of the schedule to the maintenance department (and to other departments) for consideration. The proposed variation is then modified based on the comments received from these departments, i.e. a new variation is designed which is once again sent to different departments for their consideration, etc. This procedure comes to an end when agreement has finally been reached between the airline schedule design department and all of the other airline departments.

When the maintenance department evaluates the proposed flight schedule variation, it checks first of all whether the schedule fulfills maintenance requirements regarding time and the airport where individual aircraft components are checked and replaced. It is also possible to calculate approximate maintenance costs for different airline schedule variations.

As cited by Etschmaier and Rothstein,[36] the maintenance department should, within the scope of the air carrier's long term plans, be sent at least general variations of the airline schedule for consideration in order to assure the timely development of a specific aircraft maintenance policy or the timely planning of needed maintenance facilities. In discussing the influence of the maintenance system on airline schedule design, Etschmaier and Rothstein pointed out the connection between airline schedules, the maintenance policy and needed maintenance facilities. To this effect, the aircraft maintenance policy together with the proposed airline schedule variations have an important influence on the maintenance facilities which are needed. On the other hand, existing facilities can to a certain extent influence the maintenance policy and airline schedules.

It has already been mentioned that many air carriers first work out one airline schedule variation that is then sent to the maintenance department for consideration.

When working out this schedule, or aircraft routing plan, the maintenance department's requirements are most often not taken into consideration. These requirements are built into the aircraft routing when the first proposal is modified. In certain cases only minor changes are needed to the proposed airline schedule, or aircraft routing, in order to satisfy the maintenance department's requirements, too. In other cases radical changes must be made to the airline schedule in order to meet the maintenance department's demands. We repeat that an aircraft routing can be made so that the first airline schedule already takes the maintenance department's demands into consideration. Richter, [128], [129] Labombarda and Nicoletti [84] and Kayukawa [76] designed aircraft routing plans that also took maintenance department demands into consideration. Richter used combinatorial programming (branch-and-bound) to solve this problem while Labombarda and Nicoletti based their model on dynamic programming.

In addition to looking into needed facilities at the main air carrier base, discussions of the influence of airline schedules on needed maintenance facilities should also establish facilities which are needed (equipment and technical teams) at individual airports where certain maintenance operations are planned to be made. The airline schedule design should also take into consideration needed manpower and equipment, and appropriate operations when servicing the airplane (loading, unloading, washing, cleaning, fueling up, etc.) at all airports in the network. Babic [8] considered both how to determine the necessary number of men and suitable equipment, and how to determine the team's optimal operation when servicing aircraft at airports.

The necessary amount of aircraft spare parts at individual airports in the network must also be established. Special estimates should not be made of the supply level of certain more expensive parts that rarely break down. Air carriers most often keep only one such part at individual airports. Should this part be put into an aircraft, word can be quickly sent to the technical base so that a new part is sent out on the first possible flight. Etschmaier and Rothstein [36] pointed out that the demand for more spare parts at airports is proportional to the number of take offs and landings at that airport. Both the failure rate and the proposed airline schedule should be taken into consideration when estimating the level of spare parts needed at individual airports.

3.10 Daily Operational Aircraft Assignment to the Planned Route

As already mentioned, the maintenance department's requirements must be taken into consideration when designing the airline schedule, i.e aircraft routing. We indicated that it is possible to compile an initial aircraft routing plan without consulting the maintenance department. The next step is to build the maintenance department's requirements into the airline schedule by coordinating the maintenance department with the airline schedule design department. It was also stressed that the maintenance department's requirements can be taken as operational constraints which must be taken into consideration when developing an aircraft routing plan.

An aircraft routing plan is usually worked out for a season lasting three months. Airline schedules are most frequently repeated on a weekly basis, although in some cases they repeat on a daily basis. The airline schedule, or routing plan, is usually made several months in advance. In the meantime, certain changes often appear in flights to be made. Operational-daily aircraft routings frequently differ from those planned due to traffic disturbances (cancelled flights and delayed flights) resulting from meteorological, technical and other factors. In most cases, dispatchers in air carrier traffic centers learn of the daily routing plan just before it starts. The cancellation of one or more aircraft due to technical reasons, when there are no reserve aircraft, results in the cancellation of certain flights since the planned schedule cannot be flown with the number of aircraft available. More will be said later on concerning the optimal tactics to organize traffic when disturbances occur in the airline schedule.

Let us now discuss the following situation. Let the flights to be made and their routing be known. The traffic operation center dispatcher has the daily task of assigning available aircraft from the fleet to individual routes. It is obvious that the aircraft can be assigned to the routes in different ways. In addition, the number of hours is known that each aircraft can fly before going for a mandatory technical checkup and parts replacement. Air carriers most often have a technical base where this work is done. It is certain that air carriers are interested in keeping aircraft in service as long as possible before they go to the technical base for repairs. The reasons for this policy are primarily economic as maintenance costs on the average make up 20-30% of the air carrier's total costs.

Dispatchers in traffic operation centers encounter the following practical problem on a daily basis: Assign the available aircraft from the fleet to specific routes so that the aircraft are kept in normal operation for as long as possible before going to the technical base.

The set of routes to which aircraft must be assigned changes daily and is normally known only several hours before the assignments must be made. This aircraft assignment to routes is combinatorial by nature. Although seasoned dispatchers can make these assignments rather successfully, it is clear that the combinatorial nature of the problem requires the utilization of certain models and modern computer techniques in order to achieve the best possible effects. The problem of assigning aircraft to routes was considered by Lee and Reno[88], Konig,[80] and Holst and Sorensen[65].

Input data in the model by Lee and Reno included the number of operating hours remaining for each aircraft in the fleet before it had to head for the technical base. Taking these hours into consideration along with the average daily utilization per aircraft and the maximum allowed time that an aircraft can be in operation between two consecutive stops at the technical base, Lee and Reno[88] calculated the latest date for each aircraft in the fleet to be in operation before going to the base. The authors obtained this data after solving the problem whose objective function is the total number of unused hours that aircraft can still fly. (Aircraft do not remain in operation until the last possible moment, since this is not feasible due to the existing airline schedule and the existence of primarily one or two technical bases). Lee and Reno [88] focussed on minimizing the total number of unused hours that all aircraft from the fleet could still spend in the air. The model developed still does not refer to the operational assignment of aircraft to routes since the model operates with average daily aircraft flying time. Also, the model does not assume the possibility of certain changes in the set of routes (cancellation of some flights, introduction of new flights, etc.).

Konig [80] used combinatorial programming (branch-and-bound) to assign aircraft to routes. He divided all aircraft in the fleet into four groups: the first group included aircraft that could remain in use for considerably longer than one day; the second group contained aircraft that could remain in use a little more than one day; the third group of aircraft could remain in use for one more day and the fourth group of aircraft could only be used for several hours more. Konig [80] also focussed on minimizing the total number of unused time that aircraft could still spend in operations. With this model, aircraft that can stay in operations for a longer time are assigned to longer routes, and vice versa. The model output provides aircraft route assignments. If the dispatcher is not satisfied with the assignment, he can make certain changes in routes, run the program again and obtain a new assignment. The procedure is repeated until the dispatcher - decision maker is satisfied with the aircraft routing assignment.

Dispatchers who assign aircraft to routes will be more willing to accept the aid of a computer in the assignment process if the human role in applying the computer program is stressed. This includes applying suitable interactive programs in assigning aircraft to routes. Holst and Sorensen [65] developed a heuristic algorithm for aircraft routing assignment and corresponding interactive computer programs. We will go into a bit more detail of the Holst-Sorensen algorithm as there is considerable justification in developing a heuristic algorithm to solve this problem, primarily due to the fact that models based on mathematical programming can require a great deal of computer run time, especially for medium and large-sized problems.

Holst and Sorensen [65] first of all approached route construction, i.e. developing an aircraft routing plan. Previous discussions treated the problem of route construction, mentioning the possibility of the FIFO or LIFO (last in first out) principle. When routes have been established, the next step is to assign aircraft to the routes. This assignment can be shown in the form of a bipartite graph (Figure 3.38).

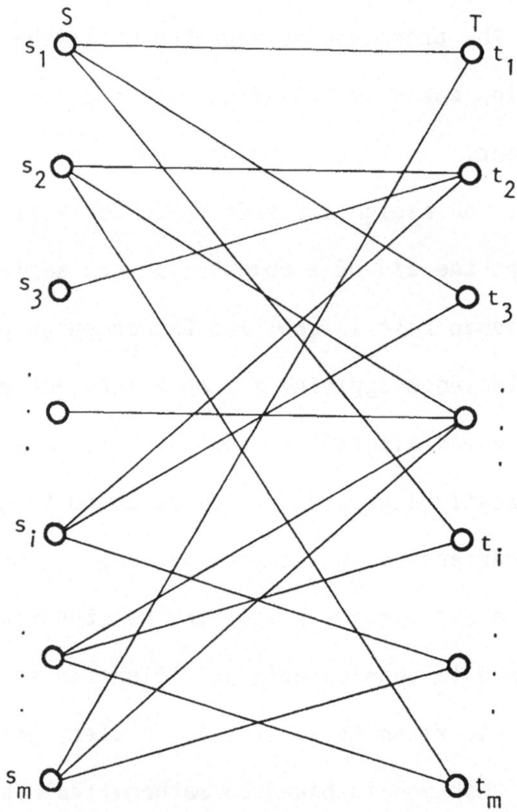

FIGURE 3.38 Assigning aircraft to routes

A specific aircraft can be assigned to a specific route only if it has "slept" at the airport where the route starts the following morning. Pairing up aircraft with routes also entails certain costs. For example, pairing node (aircraft) S_1 with node (route) t_3 entails cost c_{13}. These costs include the unused flight hours when aircraft s_1 completes the flights along route t_3 and goes to the technical base for work even though this aircraft can still fly for a certain time (Fig. 3.39).

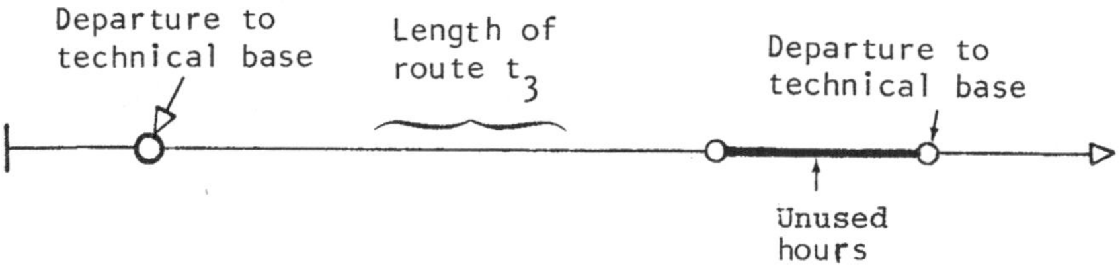

FIGURE 3.39 Utilization of possible aircraft flying time

The objective function represents the total number of unused hours between two technical checkups for all aircraft from the fleet. It is clear that the focus is on minimizing the total number of unused flight time . As mentioned by Holst and Sorensen, it might happen that no feasible solution exists or that the total number of unused time is very large. Such is the case when an aircraft has to go for maintenance work within 24 hours and none of the routes starting from the airport where the aircraft is located ends at the airport where the technical base is located. Holst and Sorensen's heuristic algorithm is comprised of changes to those routes that start from the airport where the aircraft is located. An example of route changes is given in Fig. 3.40.

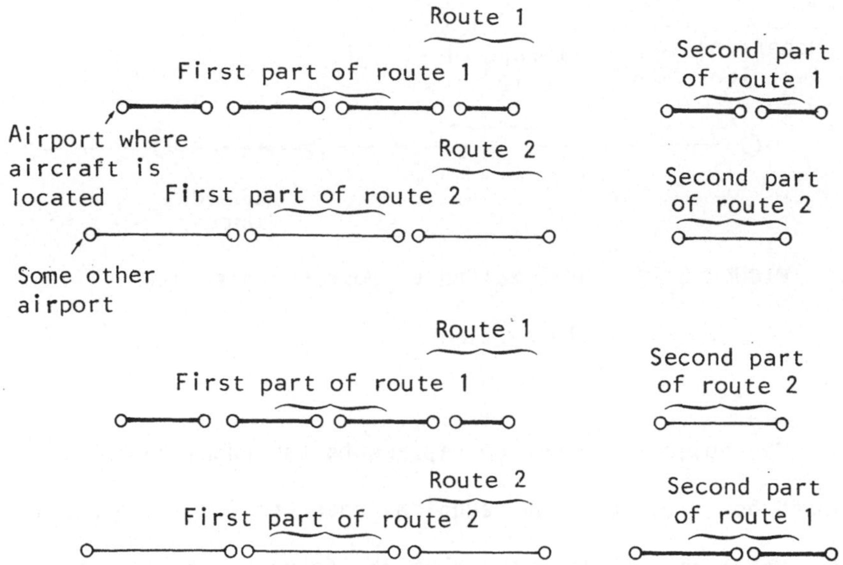

FIGURE 3.40 Changes in route parts

In the example shown above, the original plan called for the first aircraft to fly along the first route, and the second to fly along the second route. This would not satisfy the maintenance department's requirements regarding the first aircraft. For this reason, the second part of route 2 was added to the first aircraft and the second part of route 1 was given to the second aircraft. For this exchange to be implemented, these two aircraft have to meet at one of the airports in the network. In the above example, this was the airport where the first parts of the original routes ended and the second parts began. Also, these routes must start at different airports. Changes were made to the original routes in this case because the first aircraft was due for maintenance work at the technical base, and the route it had been assigned did not end at the airport where the technical base was located. In order to change route parts, the second aircraft must not be due for maintenance work during the following day. Changes in route parts continue until a feasible solution is obtained that satisfies the traffic planner.

As mentioned by Holst and Sorensen,[65] computer programs developed are in use on the SAS DC-9 fleet network. Observation was made of 559 flights to be made. Aircraft landed at 46 different airports and spent the night at 12 airports in the network. The SAS DC-9 fleet contains 32 aircraft. Very little computer time was needed both in generating the original routes and in their later modification.

This heuristic algorithm is very simple and conveniently applied, both because of the small amount of computer time and because the dispatcher is the main decision maker. This latter fact has made models of this type considerably more acceptable than models which reduce the dispatcher's role.

3.11 Airline Schedule Perturbations

Our previous discussions of airline scheduling problems demonstrated that designing an airline schedule is an extremely complicated process. We also discerned that the airline schedule has an essential effect on both the air carrier's economic results and on the quality of the transportation services provided. However, the manner in which a planned airline schedule is executed also has a considerable effect on the air carrier's economic results and on the quality of transportation services. Perturbations in carrying out the schedule are everyday occurrences and there are numerous factors behind them. For example, cancelled or delayed flights can be caused by meteorological reasons, technical reasons, late or absent crew members, poorly calculated block times and service times at certain airports, etc. Very few works have been devoted to the problem of perturbations in executing a planned airline schedule. Different aspects of this problem were treated by Etschmaier and Rothstein,[35] and Teodorovic and Guberinic[157]. When one or several aircraft are taken out of operations for technical reasons, the air carrier can use one of the following four strategies to remove the disturbance :

I introduce one or more reserve aircraft into operations should there be any available,

II design a new airline schedule so that none of the planned flights is cancelled, but with a certain number of delayed flights,

III cancel a certain number of planned flights, with none of the remaining flights being delayed, and

IV cancel a certain number of planned flights with a certain number of the remaining flights being delayed.

Smaller air carriers usually cannot apply strategy I since they most often do not have reserve aircraft available. For strategy II, a large number of flights are very late since none of the planned flights is cancelled and operations are run with a smaller number of aircraft. In some situations, large delays can carry airline schedule perturbations over into the following day. Clearly, the level of passenger service is very low in this case. Strategies III and IV are linked to additional air carrier costs since in most cases hotel lodgings must be paid for passengers whose flights have been cancelled. The air carrier's planned revenues are also considerably reduced when applying these strategies since a certain number of planned flights have been cancelled.

3.12 Designing a New Airline Schedule to Minimize Total Delays When an Aircraft Breaks Down

Let us consider operational strategy II. In this case, a new airline schedule is designed so that none of the planned flights is cancelled, although a certain number of flights are delayed.

The following model was developed by Teodorovic and Guberinic[157].

Let us assume that all aircraft on the network have the same carrying capacity.

We denote by $x_1, x_2, \ldots, x_k$, respectively, flights to be flown. For each flight x_i we know planned departure time a_i and planned time b_i in which the aircraft is ready to make a new flight after completion of flight x_i . Time interval $b_i - a_i$ is the sum of the time the aircraft needs to complete flight x_i and the time needed to prepare the aircraft for a new flight after completion of flight x_i. We also know the number of passengers p_i on the aircraft during flight x_i .

When designing a new airline schedule, the following condition must be satisfied : no aircraft can take off earlier than planned by the old airline schedule which has been perturbed.

In other words, if we denote departure time on flight x_i by c_i, the new airline schedule must satisfy the following :

$$c_i \geqslant a_i, \qquad i = 1,2,\dots,k \tag{3.37}$$

Departure time by the new airline schedule is denoted by c_i. We denote by d_i the moment in the new airline schedule when the aircraft is ready for a new flight after completion of flight x_i. It is clear that :

$$d_i = c_i + (b_i - a_i), \qquad i = 1,2,\dots,k \tag{3.38}$$

Delay e_i of flight x_i is :

$$e_i = c_i - a_i, \qquad i = 1,2,\dots,k \tag{3.39}$$

We must also bear in mind the number of passengers p_i on flight x_i. To this effect, it is not the same if there are 30 or 100 passengers with a 2-hour delay.

Total delay on flight x_i is :

$$w_i = e_i p_i, \qquad i = 1,2,\dots,k \tag{3.40}$$

Total delay on the entire network is :

$$W = \sum_{i=1}^{k} w_i \tag{3.41}$$

Since we want to minimize total delay on the entire network, the problem can be formulated as follows : Determine departure times $c_1, c_2, \dots, c_k$ of flights $x_1, x_2, \dots, x_k$ so that :

$$W = \sum_{i=1}^{k} (c_i - a_i) \cdot p_i \rightarrow min \tag{3.42}$$

whereby :

$$c_i \geqslant a_i, \quad i = 1,2,\ldots,k \tag{3.43}$$

Flights x_i represent nodes in the network $G(N,A)$ shown on Figure 3.41. In other words :

$x_i \in N, \quad i = 1,2,\ldots,k$

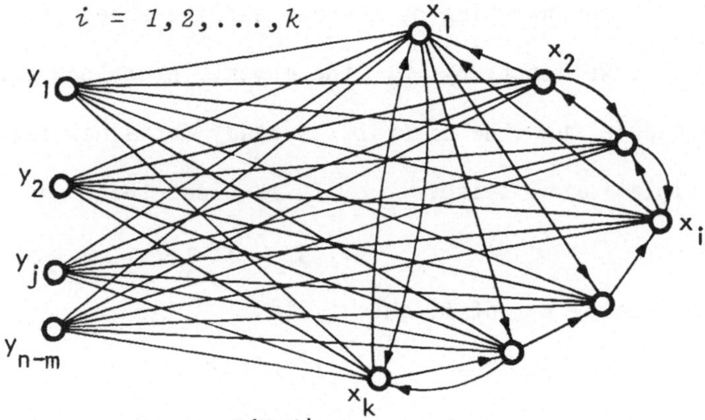

FIGURE 3.41 Graph $G(N,A)$

We denote by n the number of aircraft that the old airline schedule was designed for, and by m the number of aircraft that have been cancelled due to technical reasons. The number of aircraft remaining is $n-m$. We denote artificial flights respectively by $y_1, y_2, \ldots, y_{n-m}$ and assume that each of the remaining aircraft must make one artificial flight after which the aircraft will no longer be in operation. The set of nodes N of network $G(N,A)$ therefore also includes nodes y_j , i.e. :

$y_j \in N, \quad j = 1,2,\ldots,n-m$

The set of branches A of network $G(NA)$ includes branches (x_i, x_j) oriented from node x_i to node x_j , if flight

x_i finishes in the city where flight x_j begins. Let the set of branches A also include nonoriented branches (x_i, y_j) linking all nodes x_i with all nodes y_j.

Flights of the remaining $n - m$ aircraft can be represented by some chain q in graph G :

$$q = \left[(x_a, x_b), (x_b, x_c), \ldots, (x_r, y_1), (y_1, x_h), \ldots, (x_p, x_q), (x_q, y_j)\right]$$

It is clear that different chains in graph $G(N, A)$ can represent different combinations of flights made by individual aircraft. Each of these chains always ends in one of the y_j nodes since each aircraft must always make one hypothetical flight at the end.

We denote by $d(x_i, x_j)$ the length of branch (x_i, x_j) This length comprises the total delay of the passengers on flight x_j made after flight x_i.

Since nodes y_j are artificial, the length of all branches (x_i, y_j) and (y_j, x_i) equal zero.

Total delay on the entire network W actually equals the length of chain q. This means that :

$$W = \sum_{(x_i, x_j) \in q} d(x_i, x_j) + \sum_{(x_i, y_j) \in q} d(x_i, y_j) + \sum_{(y_i, x_j) \in q} d(y_i, x_j) \qquad (3.34)$$

We want to find at least one (there may be more) of the chains in graph $G(N,A)$ with the least total length, with each of nodes x_i and y_j included in the chain only once.

This problem can be solved using combinatorial programming (the branch-and-bound method).

Passenger delay on flight x_j made after flight x_i is :

$$d(x_i, x_j) = \begin{cases} 0, & d_i \leq a_j \\ (d_i - a_j)p_j, & d_i > a_j \end{cases} \qquad (3.45)$$

We also have :

$$d(x_i, y_j) = d(y_j, x_i) = d(x_o, x_i) = 0,$$

$$i = 1, 2, \ldots, k \qquad j = 1, 2, \ldots, n-m \qquad (3.46)$$

We take any part of chain $q, x_o, \ldots, x_i$ or $x_o, \ldots, y_j$ and join it to a number g_{s,x_i} or g_{s,y_j} representing the lenght of all branches contained in the section of the chain under observation. Index s represents the number of nodes contained in the chain section, and index x_i or y_j is the last node in the chain section under observation.

It is clear that :

$$g_{s,y_s} = g_{s+1,x_k} \qquad (3.47)$$

$$g_{1,x_i} = 0, \qquad i = 1, 2, \ldots, k \qquad (3.48)$$

The branching procedure comprises two main steps. The first is choosing the subset of the chain to be branched, while the second step comprises branching the chosen subset.

The chain subset chosen for branching satisfies this equation :

$$g_{(s,x_i)^*} = \min \{g_{s,x_i}\} \qquad (3.49)$$

Index i regards all candidates for further branching. The new chain subset contains chains with nodes from the old subset (before branching) and new nodes from the neighboring last node in the chain (Figure 3.42). Nodes already included in the chain cannot be included in the chain a second time.

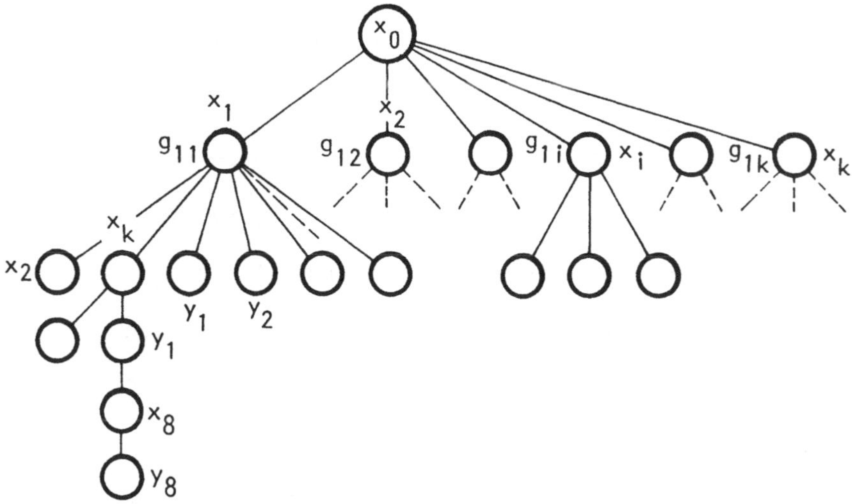

FIGURE 3.42 Branch-and-bound

When a subset is chosen for further branching, it must be checked that :

$$g_{s,x_i} < z$$

where z is the length of the shortest path discovered up to the moment in which we are observing the tree being branched.

If this condition is not fulfilled, branching of the chosen subset stops.

When a feasible solution is found (a solution containing all nodes x_i and y_j), value g_{k+n-m,y_p} is compared with upper boundary z . If this value is less than z , then it becomes the new upper boundary. The procedure is completed when value g_{k+n-m,y_p} is less than the value of the upper boundary of all subsets that are candidates for branching.

Let us illustrate the procedure with the following example.

A fleet consists of 3 aircraft of the same carrying capacity. There are 8 flights to be made during the day. The space-time diagram on Figure 3.42 presents the planned flights and the planned departure times of each flight.

The three aircraft are denoted by a,b and c . Flights are denoted by $x_1,x_2,\ldots,x_8$.

Aircraft b (the heavy line) that should fly on route A_1-A_3-A_1 is removed from operations for technical reasons.

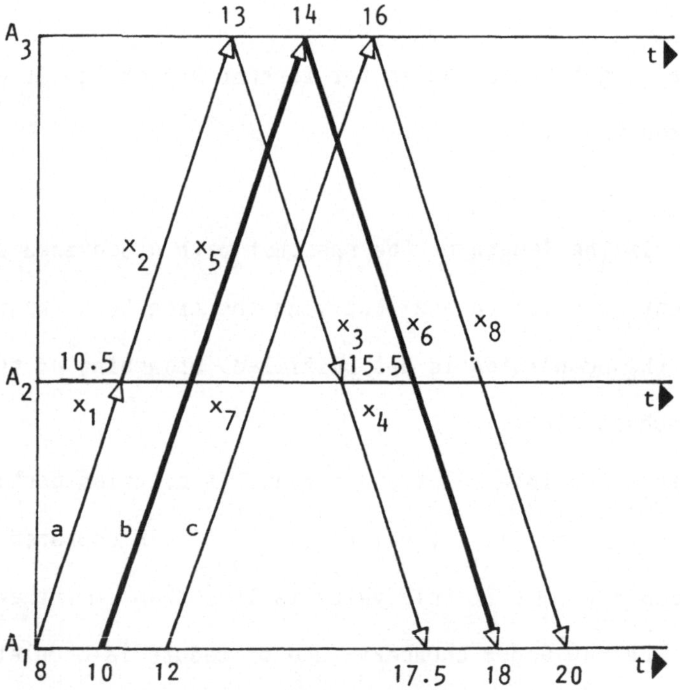

FIGURE 3.43 Space-time diagram

Design a new airline schedule to be made by aircraft a and c so that delays on the entire network are minimized.

Graph $G(N,A)$ is shown on Figure 3.44.

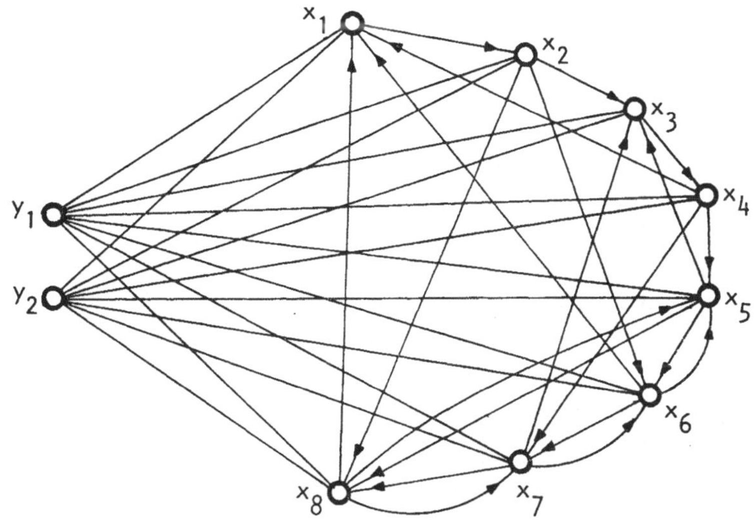

FIGURE 3.44 Graph $G(N,A)$

Planned departure times according to the old airline schedule and the number of passengers per flight are given in Table 3.4.

The optimal solution is shown on Figure 3.45.

Artificial node y_1 limited the flights to be made by aircraft a and c. Aircraft a should make flights x_1, x_2, x_3, x_4, x_7 and x_8 (in that order) and aircraft c flights x_5 and x_6. The optimal value of the objective function is 302.5 hours.

Table 3.5 gives the planned and actual departure times, and the delay on the 8 flights.

TABLE 3.4

Flight x_i	*Planned departure time* a_i	*Planned time aircraft is ready for a new flight after completion of* x_i b_i	*Number of passengers on flight* x_i p_i
x_1	8	10	105
x_2	10,5	13	25
x_3	13	13,5	35
x_4	15,5	17,5	110
x_5	10	14	65
x_6	14	18	70
x_7	12	16	25
x_8	16	20	30

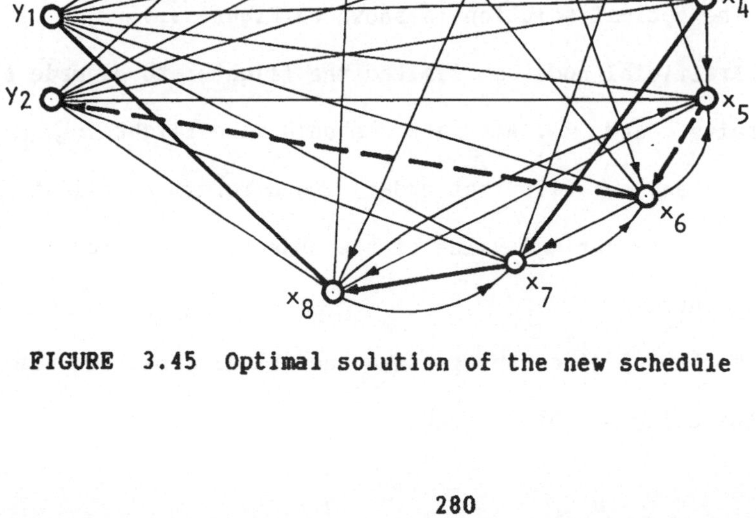

FIGURE 3.45 Optimal solution of the new schedule

TABLE 3.5

Flight x_i	*Planned departure time* a_i	*Actual departure time* c_i	*No. of passengers on flight* x_i p_i	*Delay on flight* x_i $w_i = (c_i - a_i)p_i$
x_1	8	8	105	0
x_2	10,5	10,5	25	0
x_3	13	13	35	0
x_4	15,5	15,5	110	0
x_5	10	10	65	0
x_6	14	14	70	0
x_7	12	17,5	25	137,5
x_8	16	21,5	30	165

Branch-and-bound techniques are convenient for problems of this type. However, for problems with greater dimensions, considerable computational difficulties may arise. We assumed that all aircraft in the model were of the same carrying capacity. The algorithm can be modified for the much more realistic situation of a fleet with aircraft of different capacities.

Further research into schedule perturbations should consider in greater detail the situation when a number of aircraft are taken out of operations. Different heuristic procedures could provide satisfactory solutions for larger transportation networks. Models that treat schedule perturbations are one of the important orientations in research dealing with airline schedule problems.

4. CREW SCHEDULING PROBLEMS

4.1. Crew Scheduling

Upon completion of the airline schedule design, scheduling crews and assigning them to flights are among the most important problems facing air carriers. Legal provisions in each country and internal company regulations regarding crews provide different constraints that must be satisfied when assigning crews to individual flights. For example, there are various constraints concerning the length of the crew's working time, regulated breaks between two flights, the daily number of takeoffs and landings, special payment for overtime, etc. Thus, the objective function that is used to assign crews to individual flights changes from case to case. Some air carriers strive to minimize total crew costs, some try to maximize crew utilization and minimize the number of reserve crew members, etc.

Crew scheduling, the total number of crew, their utilization and total crew costs depend to a large extent on the chosen airline schedule variation. However, although there is a significant interdependence between the airline schedule design process and the crew scheduling process, these two processes are usually separated in order to facilitate problem solving. In air transportation as in other fields of transportation, the vehicle schedule is first designed and then the crew is scheduled based on it.

Crew costs in air transportation are extremely high, comprising on the average 10-15% of total air carrier costs. Therefore, crew scheduling and assignment to routes must be given due attention. By applying models that have already been developed, modifying them and creating new crew scheduling models, considerable financial savings can be achieved compared to crew assignments made by hand without computers, using methods from practical experience. As noted by Marsten[99] and Shepardson,[99] application of a suitable crew scheduling model by Pacific Southwest Airways realized savings of 5-32% compared to the schedule designed without the aid of a computer. The same authors mention that the largest world air freighting company, Flying Tigers, with 6 B-747 aircraft among its fleet, realized savings of around $300,000 by applying a suitable crew scheduling method to its B-747's alone.

The first papers devoted to the problem of crew scheduling appeared at the beginning of the 1960s. Some of the models developed in the past two decades are in practical application, while others developed within specific airline carriers have not been published. The following discussions will consider the most important crew scheduling methods developed in the past two decades.

4.2. Formulating Crew Scheduling Problems

Crew scheduling problems can be formulated as follows : for a given airline schedule, schedule the crew so that total crew costs are minimized while simultaneously respecting all provisions that regulate crew operations so as to not endanger the security of the aircraft.

Schedules can be designed for a day, a week, etc. The literature contains numerous terms for different crew activities or duties. The following presentations will use the terminology found in Ball and Roberts.[11]

The concept of leg comprises a nonstop flight between a pair of cities characterized by the city and departure time and the city and landing time. Crew members spend a certain amount of time on the ground between two nonstop flights. This time, however, is not considered to be crew rest time. Crew rest time comprises the time spent on the ground that is greater or equal to some prescribed time interval. The series of flights made by crew members that is separated from a second series of flights by a rest time is called work duty or the crew's working time. Working time usually includes one hour before the first flight when the crew prepares for the flight, the length of the flights to be made, the time crew members spend on the ground between flights and 15-30 minutes after completion of the last flight. The maximum allowed crew working time is regulated by special provisions. Average working time usually covers several hours. The series of duties separated from each other by rests in such a way that the first flight starts from the base and the last flight ends up at the air carrier's base is called a rotation. Figure 4.1 indicates the legs of the work duty, time between flights, rest time, and rotation.

The schematic manner of presenting crew duties originated with Ball and Roberts.[11]

Rotation usually lasts several days. Each rotation is characterized by certain rotation costs. These costs are calculated in different ways by different air carriers based on the length of duty within the rotation, total flight time, the number of work duties, the time crew members spend outside their homes (rotation time), etc. Frequently crew members leaving for duty or returning from duty are found in passenger seats. For example, if the air carrier base is in city A, and the crew should start duty in city B, this crew will be on a flight from city A to city B in passenger seats. Air carriers also calculate and pay crews for the time they spend in the plane going to or returning from duty.

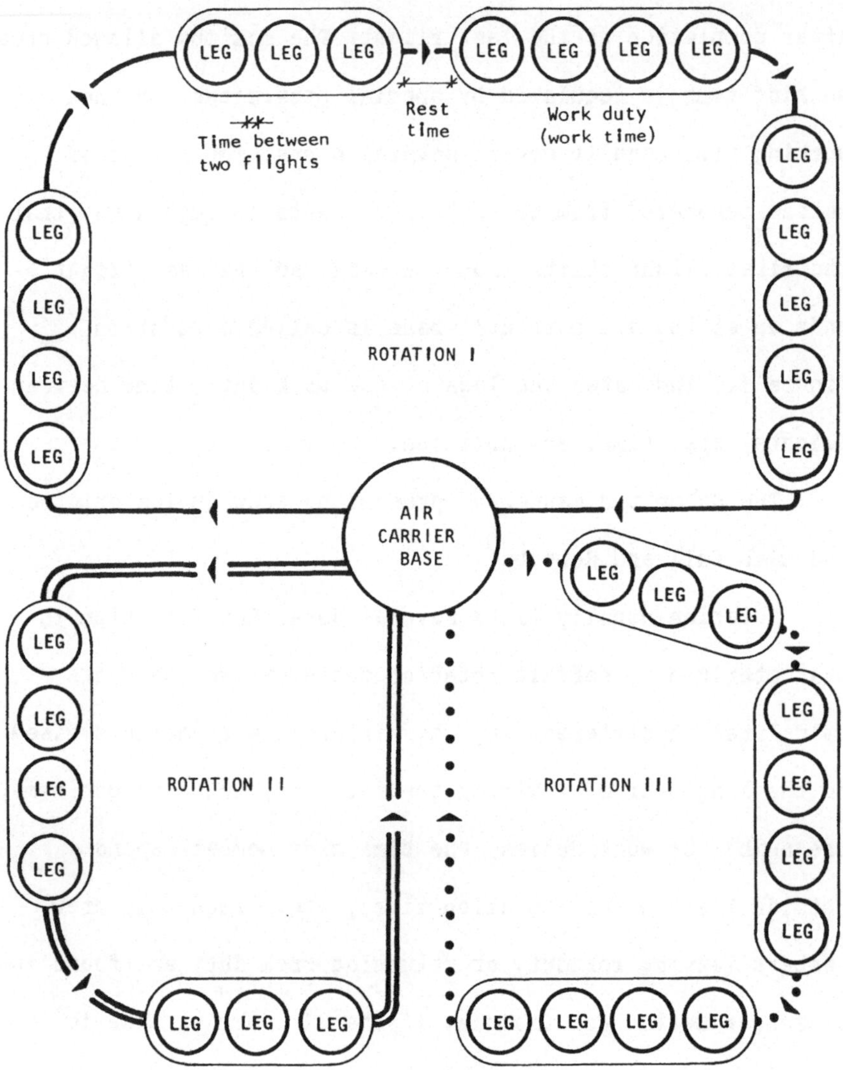

FIGURE 4.1 Legs, work duties, time between flights, rest time and rotation

We denote by :

r_j - possible rotation $(j=1,2,\ldots,n)$

n - number of possible rotations,

c_j - the cost of rotation x_j ,

m - number of flights to be made according to the given airline schedule.

The following binary variables are also introduced :

$$a_{ij} = \begin{cases} 1, & \text{if flight } i \text{ is part of rotation } j, \\ 0, & \text{otherwise} \end{cases}$$

$$x_j = \begin{cases} 1, & \text{if rotation } r_j \text{ exists in the crew schedule,} \\ 0, & \text{otherwise} \end{cases}$$

The crew scheduling problem can be formulated as follows :

$$\sum_{j=1}^{n} c_j x_j \longrightarrow min \qquad (4.1)$$

$$\sum_{j=1}^{n} a_{ij} x_j = 1 , \qquad i = 1,2,\ldots,m \qquad (4.2)$$

$$x_j = 0 \quad or \quad 1, \qquad j = 1,2,\ldots,n \qquad (4.3)$$

Expression (4.1) strives to minimize total rotation costs. Expression (4.2) stipulates that every flight $i(i=1,2,\ldots,m)$ is only within one rotation $r_j(j=1,2,\ldots,n)$, while expression (4.3) denotes the fact that each of the possible rotations can (but does not have to) participate in the final solution one time at most. The problem defined by expressions

(4.1). (4.2) and (4.3) is called a set partitioning problem or set dividing problem.

Since expression (4.2) anticipates that each flight belongs to only one rotation, we conclude that only one crew can be on the aircraft for each flight - the crew on duty for that flight. On the other hand, it was already mentioned that crew members are often on aircraft in passenger seats, going to or returning from duty. In this case it is allowed, so that flights are within the composition of one or more rotations, and that individual flights can contain one or more crews. In other words, expression (4.2) becomes an inequality in this case :

$$\sum_{j=1}^{n} a_{ij} x_j \geq 1 \quad , \qquad i = 1,2,\ldots,m \qquad (4.4)$$

We would mention that rotation costs with only one crew participating are not equal to rotation costs containing the same flights but with some flights having other crews in passenger seats as they go to or return from duty. When expression (4.2) is replaced by expression (4.4) in the problem formulation, this is called a set covering problem.

Various constraints linked to short-haul or long-haul transportation have resulted in the development of different crew scheduling models for short-haul, medium-haul and particularly long-haul transportation. We would underscore that the crew scheduling problem comprises only the choice of a certain number of rotations from the set of all possible rotations for which the air carrier has the least possible crew costs. The schedule does not include assigning individual pilots or stewardesses to specific rotations or flights. After completion of the crew scheduling plan for a certain period of time (week, month), an approach is made to assigning individual pilots and stewardesses to specific duties. This problem will be hereafter called the crew rostering problem. When rostering pilots and stewardesses to specific duties, an attempt is made to balance the utilization of individuals.

The failure of some crew member to come to work can cause delays in a large number of flights. In order to avoid this, most air carriers have reserve crews that are considered to be on work duty while being kept in reserve. When needed, different reserve crew members take the place of their colleagues who are sick or late. It is clear that the total crew costs of an air carrier depend among other things on the number of needed reserve crew members. Therefore, calculating the number of reserve crew must be given greater attention. Care must also be given to organizing crew transportation to and from flights. Some air carriers organize their crew's transportation on the way to flights. A more rational organization of this transportation can ensure both the needed crew punctuality when arriving for a flight, and the least possible transportation costs.

4.3. Crew Scheduling for Long-Haul Flights

The structure of crew costs in long-haul transportation differs from these cost structures in short-haul and medium-haul transportation. Crew work duties in long-haul transportation normally last from one to three weeks. A considerable share of crew costs is thereby composed of hotel lodgings and daily expense allowances.

Airline schedules in long-haul traffic are normally repeated every week. Our further considerations will present Mondria's[106] model which outlines the experience of KLM in solving the problem of crew scheduling for long-haul transportation with an airline schedule that is repeated on a weekly basis.

Figure 4.2. presents the long-haul airline network. Flights to be made during the week are denoted respectively by 1, 2, ..., 74. Cities with transportation between them are denoted by $A, B, C, \ldots, O$.

We introduce the assumption that crew can sleep in any of cities $A, B, C, \ldots, O$. (In some air companies, crew can only spend the night in specific cities. The choice of city in which the crews sleep influences the final solution, or

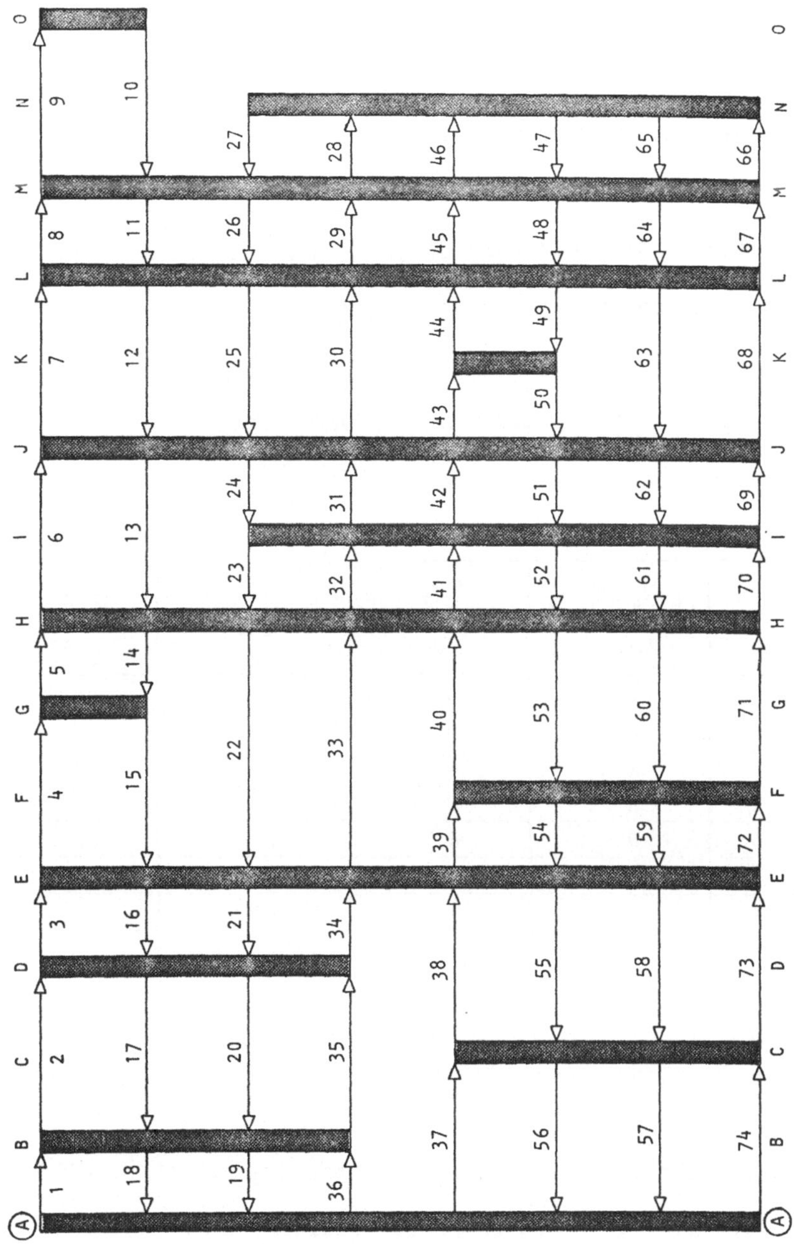

Figure 4.2 Long-haul network

the crew schedule, the needed number of crew and total costs).

Bearing in mind the legal and company constraints regarding permitted crew working time, we must first make a list of all possible work duties and calculate the corresponding costs of each one.

TABLE 4.1 List of all possible work duties

No	*City where work duty begins*	*City where work duty ends*	*Flights included in work duty*	*Work duty costs*
1	A	B	1	270
2	A	D	1,2	400
3	A	E	1,2,3	510
4	B	D	2	110
5	B	E	2,3	250
.	.	.	.	.
.	.	.	.	.
.	.	.	.	.

The above Table (Table 4.1) presents a list of all possible crew work duties on the network under observation. Work duty costs are calculated in different ways by different air carriers. However, in spite of the different ways of calculating, work duty costs are most often not equal to the sum of the costs of the flights entering into the work duty composition. In addition to work duty costs, a calculation must also be made of the costs of matching individual work duties. To this effect, a specific crew during a week of long-haul transportation carries out several work duties making up one rotation. When constructing the rotation we start with one work duty to which we will successively add the other work duties. The cost of matching two work duties is represented by overnight and daily allowance costs caused by this matching In some cases, crews spend several days in a city after completion of a work duty after which time they start a new duty. In some cases the cost of matching two work duties can be very large.

Let the air carrier's base be located in city A . We start construction of the first rotation by observing the possible work duties linked to the farthest city (city O). Let these be work duties composed of only one flight, i.e. work duty (9) and work duty (10) . (In the example taken, work duties (7, 8, 9) , (8, 9) , (10, 11) , etc. are not possible because they exceed the allowed working time). We match work duty 9 with work duty 10 . The rotation now reads :

[(9), (10)]

Let the cost of matching work duty (9) with work duty (10) be 800.

In the next step we take the next farthest city. This is city *N* . Flights coming to city *N* include flights 28, 46 and 66. Flights leaving this city include flights 27, 47 and 65. We consider the possible work duties linked to city *N* . Let this be the following duties :

(27) , (28) , (29,28) , (46) , (47) , (65) ,

(65,64) , (66) , (67,66)

The cost of matching work duties ending in city *N* with duties starting from city *N* are given on the following matrix :

	(27)	(47)	(65)	(65,64)
(28)	250	900	1100	650
(29,28)	(200)	(750)	800	(540)
(46)	(250)	(900)	1100	(650)
(66)	400	700	580	320
(67,66)	(390)	(680)	360	(310)

We can see, for example that flight 28 appears in work duty (28) and in work duty (29,28) . In the same vein, flight 65 appears in work duty (65) and in work duty (65,64) . The next step is to establish which of the work duties containing flight 28 should be matched with one of the work duties containing flight 65 so as to minimize matching costs. We see that the candidates for matching are on the one hand work duties (28) and (29,28), and on the other work duties (65) and (65,64). The least costs involve matching work duty (29,28) with work duty (65,64) (these costs are 540). Pairs of work duties corresponding to the least possible matching costs within a specific group of possible matchings are determined in the same manner.

Our further discussion will only consider those pairs of work duties entailing the smallest matching costs within a certain group. The new matching cost matrix reads :

	(27)	(47)	(65,64)
(29,28)	200	750	540
(46)	250	900	650
(67,66)	390	680	310

We want to assign each of the work duties arriving in city N ((29,28), (46), (67,66)) to exactly one of the work duties leaving city N ((27), (47), (65,64)). This type of problem in the literature is called an assignment problem. Since we want to find the optimal matchings, i.e. the optimal assignment of some work duties with others, this concerns an optimal assignment problem.

We denote by c_{ij} the costs of matching work duty i with work duty j. We introduce binary variable x_{ij} into the discussion, defined as follows :

$$x_{ij} = \begin{cases} 1, \text{ if work duty } i \text{ is matched with work duty } j, \\ 0, \text{ otherwise} \end{cases}$$

The problem of optimally assigning work duties in a specific city can be defined as follows : For a given matrix $[c_{ij}]$ of matching costs for individual work duties, determine x_{ij} for $i=1,2,\ldots,n,\ j=1,2,\ldots,n$ so that :

$$\sum_{i=1}^{n} \sum_{j=1}^{n} c_{ij}\, x_{ij} \longrightarrow min \qquad (4.4)$$

$$\sum_{i=1}^{n} x_{ij} = 1 \quad , \quad j=1,2,\ldots,n \qquad (4.5)$$

$$\sum_{j=1}^{n} x_{ij} = 1 \quad , \quad i=1,2,\ldots,n \qquad (4.6)$$

$$x_{ij} = \begin{cases} 1, \text{ if work duty } i \text{ is assigned to work duty } j \\ 0, \text{ otherwise} \end{cases}$$

If we subtract some constant r_k from all elements of the k-th row of matrix $[c_{ij}]$, we will get a new matrix $[c'_{ij}]$ that satisfies the following :

$$\sum_{i=1}^{n} \sum_{j=1}^{n} c'_{ij}\, x_{ij} = \sum_{i=1}^{n} \sum_{j=1}^{n} c_{ij}\, x_{ij} - r_k \sum_{j=1}^{n} x_{kj} =$$

$$= \sum_{i=1}^{n} \sum_{j=1}^{n} c_{ij}\, x_{ij} - r_k \qquad (4.8)$$

Based on expression (4.8) we can conclude that for the same set of variables x_{ij}, the minimum is achieved by

$$\sum_{i=1}^{n} \sum_{j=1}^{n} c_{ij} x_{ij} \quad \text{and} \quad \sum_{i=1}^{n} \sum_{j=1}^{n} c'_{ij} x_{ij}$$

We would arrive at the same conclusion if we subtract some constant s_l from the l-th column.

We denote by r_i the smallest element in the i-th row of matrix $[c_{ij}]$. For $i = 1,2,\ldots,n$ we subtract r_i from each element in the i-th row. This gives us some new matrix $[c'_{ij}]$. We denote by s_j the smallest element in the j-th column of matrix $[c'_{ij}]$ and for $j = 1,2,\ldots,n$ we take away s_j from each element in the j-th column of matrix $[c'_{ij}]$. We designate the new matrix we obtain as $[c^*_{ij}]$. All elements of reduced matrix $[c^*_{ij}]$ are nonnegative. We assume that we can choose from each row and each column exactly one element of matrix $[c^*_{ij}]$ equal to zero. Since we have chosen these elements (for them $x_{ij} = 1$ and $c^*_{ij} = 0$), this is $\sum_{i=1}^{n} \sum_{j=1}^{n} c^*_{ij} x_{ij} = 0$

This gives us the optimal assignment because

$$\sum_{i=1}^{n} \sum_{j=1}^{n} c^*_{ij} x_{ij} \geqslant 0.$$

This means that to establish the optimal work duty assignment, we must find n zeros in matrix $[c^*_{ij}]$ so that none of them is located in the same row or the same column as another. In order to find these zeros, we equal all other positive elements of matrix $[c^*_{ij}]$ with 1. This gives us new matrix $[\bar{c}_{ij}]$ with all zeros and ones. Based on the Konig-Egervary theorem (for more details see Roberts[130]) the minimum number of zeros so that none is located in the same row or column with another in matrix $[\bar{c}_{ij}]$ equals the minimal number of lines (rows or columns) that can cover all zeros in matrix $[\bar{c}_{ij}]$. Should the minimum number of lines that can cover all zeros equal n , then we have found the optimal assignment. Kuhn[83] developed the following algorithm that is known in literature as the Hungarian algorithm. We will present this algorithm in the same way as Roberts :

INPUT : Matrix $[c_{ij}]$ of dimensions $n \times n$

OUTPUT : Optimal assignment (determining the values of x_{ij} for $(i=1,2,\ldots,n, \quad j=1,2,\ldots,n)$

STEP 1

STEP 1.1 For $i=1,2,\ldots,n$ find r_i representing the smallest component in the i-th row of matrix $[c_{ij}]$

STEP 1.2 Denote by $[c'_{ij}]$ the matrix whose elements are obtained by subtracting element $r_i (i=1,2,\ldots,n)$ from all components of the i-th row of matrix $[c_{ij}]$

STEP 1.3 For $j=1,2,\ldots,n$ find s_j representing the smallest element in the j-th column of matrix $[c'_{ij}]$

STEP 1.4 Denote by $[c^*_{ij}]$ the matrix whose elements are obtained by subtracting element s_j

$j' = 1, 2, \ldots, n$ from all elements in the j-th column of matrix $[c'_{ij}]$

STEP 2

STEP 2.1 Find the smallest number of lines that can cover all zeros in matrix $[c^*_{ij}]$

STEP 2.2 If the number of lines is less than n, go to Step 3. Otherwise, go to Step 4.

STEP 3

STEP 3.1 Regarding the zeros covered by lines completed in Step 2.1, find the smallest uncovered element and denote it by m.

STEP 3.2 Subtract m from all uncovered elements in matrix $[c^*_{ij}]$. Add m to all elements of matrix $[c^*_{ij}]$ that are covered by two lines. Return to Step 2.

STEP 4

STEP 4.1 Find the set of n zeros in which none is located in the same row or column as any other.

STEP 4.2 Should component c^*_{ij} equal zero (one of the zeros), put $x_{ij} = 1$. Otherwise, put $x_{ij}=0$.

We note that in Step 3.2 the algorithm transforms the matrix in order to obtain a new one corresponding to a larger value of minimum number of lines that are able to cover all zeros. This new matrix can also be obtained by subtracting or adding some constant from different rows or columns of the previous matrix.

Let us return to matching crew work duties in city N, and apply the Hungarian algorithm. The work duty matching cost matrix in this city reads :

$[c_{ij}]=$

	27	47	65,64
29,28	200	750	540
46	250	900	650
67,66	390	680	310

Matrix $[c'_{ij}]$ reads :

$[c'_{ij}]=$

	27	47	65,64
29,28	0	550	340
46	0	650	400
67,66	80	370	0

Matrix $[c^*_{ij}]$ reads :

$[c^*_{ij}]$ =

	27	47	65,64
29,28	0	180	340
46	0	280	400
67,66	80	0	0

The least number of lines that can cover all zeros is two (the first column and the third row). Since the smallest number of lines is less than three $(n = 3)$ we subtract 180 from all uncovered elements, and add 180 to the element that is covered twice. The new matrix reads :

	27	47	65,64
29,28	0	(0)	160
46	(0)	100	220
67,66	260	0	(0)

In this case, the smallest number of lines that can cover all zeros equals three. The set of three zeros with none belonging to the same row or column as another includes the zeros that are circled in the above matrix. This means that :

$$x_{12} = 1, \quad x_{21} = 1, \quad x_{33} = 1$$

We conclude that the following matching of crew work duties is optimal :

(29,28) *with* (47)

(46) *with* (27)

(67,66) *with* (65,64)

Total matching costs for these work duties are :

$$750 + 250 + 310 = 1310$$

Now we already have four incomplete crew rotations :

[(9),(10)]

[(29,28),(47)]

[(46),(27)]

[(67,66),(65,64)]

We note that no flight existed that could link cities *O* and *N* . However, this was not the case of the next pair of cities *N* and *M*. (In constructing the crew rotation we move "backwards" from the farthest city towards the air carrier base). Flights that arrive at city *M* are flights 8, 10, 29, 47, 65 and 67. Flights departing from city *M* include flights 9, 11, 26, 28, 46, 48, 64 and 66. However, some of these flights are already included in work duties that have become part of rotations. Incoming flights that are not included in earlier discussed work duties include flights 8 and 45. Let the work duties that these flights can be part of be duties (7,8), (8) and (45). Outgoing flights that are not included in work duties include flights 11, 26 and 48. Let the work duties that these flights can be part of be duties (11), (26), (48) and (48,49) and let the cost of matching work duties that end in city *M* with work duties that start with a flight from city *M* be given in the following matrix :

	(9,10)	(46)(27)	(11)	(26)	(48)	(48,49)
(9, 10)	(10^6)	(280)	(420)	(390)	700	(560)
(29,28)(47)	(320)	(330)	(480)	(460)	750	(630)
(46) (27)	(340)	(10^6)	(530)	(490)	770	(680)
(7, 8)	200	270	400	360	680	540
(8)	(190)	(220)	(380)	(330)	500	(490)
(45)	(380)	(620)	(600)	(510)	790	(690)

Work duty (9,10) starts with a flight departing from city *M* . On the other hand, work duty (9,10) finishes with flight 10 that arrives in city *M* . Therefore, it is theoretically possible to pair work duty (9,10) with itself. In order to avoid this, such a matching will comprise extremely high matching costs that will make in impossible for a work duty to be matched with itself in the final solution.

As was the case previously, we will only consider those pairs of work duties that entail minimum matching costs within a specific group of work duties. The new matching cost matrix reads :

	(9 , 10)	(46)(27)	(11)	(26)	(48,49)
(9 , 10)	10^6	280	420	390	560
(29,28)(47)	320	330	480	460	630
(46)(27)	340	10^6	530	490	680
(8)	190	220	380	330	490
(45)	380	620	600	510	690

After application of the Hungarian algorithm, we get :

$$x_{13} = 1,\ x_{22} = 1,\ x_{34} = 1,\ x_{45} = 1,\ x_{51} = 1$$

This solution corresponds to matching costs equalling 2,110. (The solution $x_{13} = 1,\ x_{22} = 1,\ x_{31} = 1,\ x_{44} = 1,\ x_{55} = 1$ **also corresponds to matching costs totalling 2,110). We choose the first solution obtained for further operations. This solution corresponds to the following matching of crew work duties :**

(9 , 10) *with* (11)

(29,28)(47) *with* (46)(27)

(46) (27) *with* (26)

(8) *with* (48 , 49)

(45) *with* (9 , 10)

By observing the flights that start and finish in cities *O* *N* **and** *M* **the following incomplete crew rotation is obtained :**

[(45), (9,10), (11)]

[(29,28), (47), (46), (27), (26)]

[(8), (48,49)]

[(45), (9,10)]

[(67,66), (65,64)]

Starting from the farthest city towards the air carrier's base city, the next step (city *L*) would apply the same methodological procedure to construct the crew rotation. When the optimal assignment has been found for city *L*, the problem would be solved for city *K*, etc. The procedure is completed when the optimal assignment problem has been solved for the last city (city *A*). The solution obtained in this manner is not optimal, for optimal work duty matchings were found in each step and only they are taken into consideration. No other possible work duty matchings are taken into consideration, even though they might lead to a better final solution. The above procedure takes one large problem and decomposes it into a series of much smaller problems. These smaller problems are those of optimal assignment. Relatively little computer time is needed to solve optimal assignment problems.

Discussions thus far have considered the costs of matching individual work duties. The costs of the work duties themselves have not been treated. It was mentioned earlier that work duty costs do not equal the sum of the costs of the branches that are a component part of the work duty. In order to facilitate the calculation of total costs in individual phases, only the difference between the work duty costs and the sum of costs entering into the work duty composition should be added to matching costs. At the end, the sum of the costs for all branches should be added to the total costs obtained in the last step.

In the example just examined, it was assumed that the crew could overnight in any of the cities. The same problem can be considered when the crew can only sleep in certain cities in the network. It is clear that the final solution depends to a sizeable extent on the choice of cities where the crew has the possibility of spending the night. Varying possible crew sleeping spots can generate a large number of different crew rotations that differ from each other both by number of crew needed and by total costs, which can greatly facilitate decision making regarding the elaboration of a crew rotation plan.

Different air carriers can have different operational constraints that must be respected when devising a crew rotation plan. For example, some have company rules concerning the number of "turnbacks" allowed during one rotation. A "turnback" is a situation when the crew returns to a city in which it was during the same rotation, or returns back in the direction it came from. For example, in part of rotation [(29 28), (47), (46), (27), (26)] there is a turnback since flight 28 finishes in city *N* , and flight 46 that is flown later also finishes in city *N* , which means that the crew lands in city *N* twice during the same rotation. In the same vein, this crew goes to city *M* three times (after completing flights 29, 47 and 27). Should turnbacks not be allowed by company regulations, they can be avoided by allotting a very high cost to any matchings that would cause a turnback. (We prevented work duties from being paired with themselves in the same manner).

In order to increase the possibility of finding a better final solution compared to the solution found by the above procedure, a certain number of possible matchings must be considered in each step in addition to the matching that corresponds to minimum costs in that step. We will use an example to show which possible work duty matchings should be considered. The matrix of the cost of matching work duties ending in city *N* with work duties starting in city *N* reads :

	27	47	65	65,64
28	250	900	1100	650
29,28	200	750	800	540
46	250	900	1100	650
66	400	700	580	320
67,66	390	680	360	310

By applying the Hungarian method the following optimal matching is obtained :

[29,28 , 47]

[46 , 27]

[67,66 , 65,64]

The cost of this matching is 1310. Now we make all possible variations of this optimal matching and calculate corresponding costs :

Optimal matching variation	*Cost*
[29,28 , 47] [46, 27] [66, 65,64]	1320
[29,28 , 47] [46, 27] [67,66 , 65]	1360
[29,28 , 47] [46, 27] [66, 65]	1580

Optimal matching variation	*Cost*
[28, 47] [46, 27] [67,66 , 65,64]	1460
[28, 47] [46, 27] [66, 65,64]	1470
[28, 47] [46, 27] [67,66 , 65]	1510
[28, 47] [46, 27] [66, 65]	1730

The optimal assignment variations do not include any flight not contained in the optimal solution. For example, the optimal solution has work duty (47) taking place after work duty (29, 28). This means that flight (47) takes place after flight (28) in the optimal solution. This order is maintained in the optimal solution variations as well. In other words, work duty (47) can appear in the optimal solution variations after work duty (28) or after work duty (29, 28) but never after work duty (46), work duty (66) or work duty (67, 66).

There are several other possible work duty matchings in each city in addition to the optimal one. For example, these matchings are also possible for city *N* :

[(28), (27)]

[(46), (47)]

[(66), (65)]

Matching costs equal 250+900+580=1730. Clearly, variations can be made of this and other possible matchings. As we have seen, it is possible to generate a large number of different matchings and their variations in each step, thereby rapidly increasing the dimensions of the problem which is combinatorial by nature. Mondria,[106] whose basic results were briefly presented in previous discussions, established that if only the variations of the optimal matching are considered, results do not disagree greatly from the optimal solution.

The choice of rotation entailing the least possible costs can also be made using the branch-and-bound method. This method was discussed in more detail in the section on airline schedule design. It is clear that if the decision maker is satisfied with the solution received without finding variations of individual optimal assignments, there is no need to apply the method of branch-and-bound.

4.4. Crew Assignments in Short-Haul and Medium-Haul Traffic as a Set Partitioning Problem

In short-haul and medium-haul traffic, crews execute work duties during the day and usually return to the air carrier's base at night. Average ground time between two work duties is shorter than in long-haul traffic. Hotel lodgings make up a smaller share of crew costs than in long-haul traffic. The biggest difference between long-haul traffic on the one hand and medium- and short-haul on the other is the total number of flights to be flown. As a rule, air carrier networks in short- and medium-haul traffic are far denser than in long-haul traffic, thereby making the dimensions of the crew work duty assignment considerably larger.

The problem of planning crew assignments and related problems have been treated in many works in the past two decades. Among others, this problem was studied by Griesshaber,[59] Arabeyre et al,[5] Pierce,[112] Littlewood,[92] Lemke et al,[89] Pierce and Lasky,[114] Rubin,[136] Marsten,[97] Christofides and Korman,[20] Etcheberry,[33] Gerbracht,[49] Marsten et al,[98] Griesshaber,[61] Marsten and Shepardson,[99] Odier et al,[111] Gerbracht,[52] Baker et al,[9] Darby-Dowman and Mitra,[21] and Ball and Roberts.[11]

The majority of these works, some of which are being applied by air carriers, approach the problem by generating all possible rotations, i.e. by designing a matrix whose rows represent flights to be flown and columns are possible rotations. Possible rotations are constructed within a specific time period, usually one day or one week. Input data include the airline schedule for the aircraft type, and relevant company and legal provisions regarding the length of working time, flight time, the number of takeoffs and landings during work hours and the length of the crew's daily rest time. Many air carriers have developed computer programs to generate rotations. When constructing the rotations, individual flights are first linked into a series of flights so that the crew does not have the possibility of changing aircraft until this series is completed. These series of flights can be made in only one way. After this phase, a smaller number of series of flights is obtained than the initial number of flights. This is particularly important in order to reduce the size of the problem. The series of flights in which the aircraft cannot be changed are linked into individual work duties, and these work duties become part of the rotation. One of the best-known computer programs to construct rotations is the CAPS program written by Robert Gerbracht for Continental Airlines. In some companies, a complete matrix of rotations is first generated, and then certain rotations are thrown out based on set criteria in order to reduce the dimensions of the problem and the matrix. As noted by Marsten et al,[98] when constructing a rotation consideration is taken of : the preset cities from which trips begin and end, the smallest and largest number of flight hours within a work duty, the number of allowed aircraft changes, the minimum time needed to change aircraft, minimum and maximum rest time between two work duties, the cities in which the crew can sleep, the maximum allowed number of work duties within a rotation, etc. When constructing possible rotations, i.e. when choosing rotations that enter into further discussions, the approach is to choose a set of rotations in which all planned flights are covered exactly once and correspond to the least possible costs (individual rotation costs must first be calculated). As was mentioned earlier, the choice of this rotation set is actually a set partitioning problem defined by expressions (4.1). (4.2) and (4.3). One of the best-known programs to solve set partitioning problems is SETPAR written by Roy

Marsten[97] which is based on a model he developed. The SETPAR program makes it possible to find one or several optimal and/or near optimal solutions. This considerably facilitates the crew planner's final choice of rotation. In any case, it is very convenient to generate several possible solutions as some other criteria might influence the choice of a solution in addition to costs.

The matrix can be reduced in several different ways. For example, BEA Company reduced the matrix almost two decades ago when striving to minimize the number of crews (all rotation costs equal one) in the following manner : every column (rotation) is removed from the matrix that is covered only by flights already covered by other rotations. As noted by Arabeyre et al,[5] this enables an initial rotation matrix with 250 rows and 25,000 columns to be reduced to a matrix with 250 rows and 3,000 columns.

Let us first discuss the model by Pierce[112] and Pierce and Lasky[114] that solves the problem of elaborating a crew assignment corresponding to the least possible costs. First the matrix must have rows that represent flights and columns that are generated rotations and must be organized so that the set of rotations corresponding to least possible costs can be easily found. The idea of how to organize the rotation matrix was first developed by Pierce[112] and then used with certain modifications in the works of Pierce and Lasky,[114] Marsten,[97]

Christofides and Korman,[20] Marsten et al,[98] and Marsten and Shepardson.[99] Before we present this idea, let us denote our problem once more :

$$\sum_{j=1}^{n} c_j x_j \rightarrow min \tag{4.5}$$

$$\sum_{j=1}^{n} a_{ij} x_j = 1 \text{ , for } i = 1,2,\ldots,m \tag{4.6}$$

$$x_j = 0 \quad \text{or} \quad 1, \text{for } j = 1,2,\ldots,n \tag{4.7}$$

where :

$$a_{ij} = \begin{cases} 1 \text{ , if flight } i \text{ is part of rotation } j \text{ ,} \\ 0 \text{ , otherwise} \end{cases}$$

Let the rotations be generated in the previous phase and let the number of possible rotations be reduced as much as possible. Matrix $[a_{ij}]$ reads :

COLUMNS (ROTATIONS)

$$\begin{array}{cc|ccccc|}
 & \text{ROWS (FLIGHTS)} & 1 & 2 & 3 & 4 & \ldots & n \\
 & 1 & 0 & 1 & 0 & 1 & \ldots & 1 \\
[a_{ij}] = & 2 & 1 & 1 & 0 & 0 & \ldots & 0 \\
 & 3 & 0 & 1 & 0 & 1 & \ldots & 1 \\
 & \cdot & \cdot & \cdot & \cdot & \cdot & & \cdot \\
 & \cdot & \cdot & \cdot & \cdot & \cdot & & \cdot \\
 & \cdot & \cdot & \cdot & \cdot & \cdot & & \cdot \\
 & m & 0 & 0 & 1 & 0 & \ldots & 0
\end{array}$$

Pierce[112] proposed that the set of all columns be divided into m subsets that exclude each other. We denote them respectively by $S_1, S_2, \ldots, S_m$. Let all columns (rotations) containing the first flight belong to set S_1, all rotations containing the second flight belong to set S_2, etc. Sets $S_1, S_2, \ldots, S_m$ are called blocks. Since all columns containing the first flight belong to block S_1, all elements in the first row of this block equal 1, while other elements in other rows equal 0 or 1. The same holds true for the second block. All elements in this row will equal 1 since this block includes a rotation containing the second flight. Other elements in the second block will equal 0 or 1 except the elements in the first row that must all equal 0 (elements in the first row of the second block must equal zero since all rotations containing the first flight are part of the first block). Elements of the first two rows in the third block equal zero, elements of the third row equal 1, others equal 0 or 1, etc. Matrix $[a_{ij}]$ obtained has a typical stairway look:

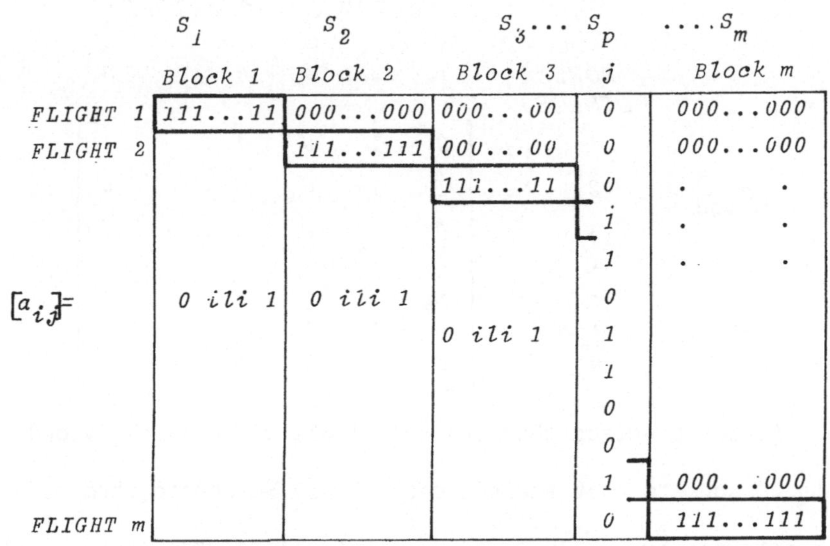

Pierce[112] took this block formation of the matrix and before applying combinatorial programming to solve the problem denoted by expressions (4.5), (4.6) and (4.7), he proposed that each rotation be lined up by increasing costs per flight covered by the rotation. In order to make this clearer, we note some rotation j in block S_p . Value

$$h_j = \sum_{i=1}^{m} a_{ij} \tag{4.8}$$

represents the number of flights covered by rotation j . Costs per flight covered by rotation j equal:

$$\frac{c_j}{h_j}$$

The first column in some block S_p should be the k-th column, and the last column in block S_p should be the l-th column, so that :

$$\frac{c_k}{h_k} \leqslant \frac{c_{k+1}}{h_{k+1}} \leqslant \frac{c_{k+2}}{h_{k+2}} \leqslant \dots \leqslant \frac{c_l}{h_l} \tag{4.9}$$

Figure 4.3 presents in tree form all possible solutions to the problem defined by expressions (4.5), (4.6) and (4.7). Rotation (column) 1 is the first column in the first block, rotation column 2 is the second column in the first block, etc. A suitable binary variable must be chosen at each level on the tree shown in Figure 4.3. For example, $x_1 = 1$ means that the first rotation is part of the final solution, $x_2=0$

that the second rotation is not part of the final solution, etc.

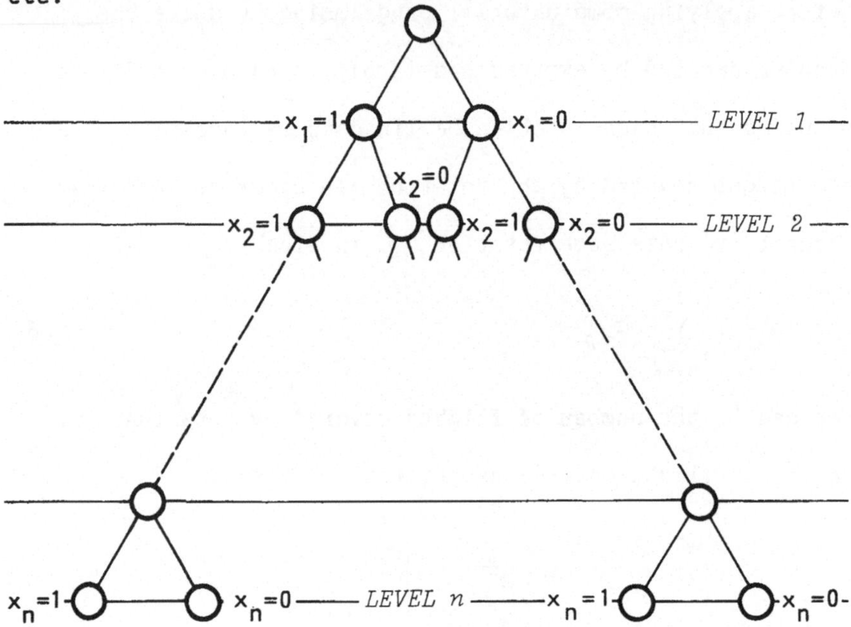

FIGURE 4.3 Possible solutions to the crew scheduling problem shown in the shape of a tree

Expression (4.6) that reads :

$$\sum_{j=1}^{n} a_{ij} \; x_j = 1 \; , \qquad i = 1,2,\ldots,m$$

can be written in the form :

$$\sum_{j=1}^{n} x_j \; \underline{A}_j = \underline{R} \tag{4.10}$$

where : $\underline{A}_j = (a_{1j}, \; a_{2j},\ldots,a_{mj})$

$\underline{R} = (r_1, \; r_2,\ldots,r_m)$

are vectors that have m elements, with all elements of vector $\underline{R}$ equal to 1.

As can be seen from Figure 4.3, there are 2^n potential solutions to our problem since each path from the first node to some node on the last n -th level represents a potential solution. We are clearly endeavoring to examine the smallest number of potential solutions possible before finding the optimal one. During our search for the optimal solution, some potential solutions must be thrown out since they are not feasible, while others should be eliminated from further consideration because that have a higher objective function value than the solution obtained that is thus far optimal.

Let us observe some level s that we have reached in our search for the optimal solution. On level s we must choose the value of binary variable x_s . The values of binary variables $x_1, x_2, \ldots, x_{s-1}$ are already defined up to our arrival at level s , and a partial solution has already been determined ($\bar{x}_1, \bar{x}_2, \ldots, \bar{x}_{s-1}$) where ($\bar{x}_1, \bar{x}_2, \ldots, \bar{x}_{s-1}$) are values taken by variables $x_1, x_2, \ldots, x_{s-1}$. We denote by $\underline{R}_s$ vector :

$$\underline{R}_s = \underline{R} - \sum_{j=1}^{s-1} x_j \underline{A}_j \qquad (4.11)$$

comprising elements $\underline{R}_s = (r_1^s, r_2^s, \ldots, r_m^s)$ that are obtained before arrival at the s -th level.

If, for example, $r_1^s=0$, this means that upon arrival at the s -th level, the first flight is already covered by some rotation and it cannot be covered by any other rotation to the end. If, for example, $r_7^s=1$, this means that in passing through the first $s-1$ levels, the seventh flight is not covered by any rotation and it can be covered by one of the remaining rotations when deciding on the s -th, $(s+1)$-th, $(s+2)$ -th,..., or n -th level.

It is clear that the following must be fulfilled for every k :

$$r_k^s \geqslant r_k^{s+1} \geqslant \dots \geqslant r_k^n \qquad (4.12)$$

We note some flight k that is covered by rotation (column) $s(a_{ks}=1)$. Let us be located on the s level, i.e. let us have to determine the value of variable x_s . If $r_k^s = 0$, this means that the k -th flight was already covered earlier by some rotation, which means that $\bar{x}_s=0$, so the k -th flight will not be covered twice. We can also conclude that in this case paths containing branch $\bar{x}_s=1$ should no longer be considered.

Let us explain how to eliminate from further consideration those paths that are not feasible, i.e. how to accelerate the process of finding the optimal solution. On level s, we have determined the value of variable x_s, which we will call $\bar{x}_s$. By determining the value of variable x_s, we know all elements of vector $\underline{R}_{s+1}$. We denote by t the smallest index i for which :

$$r_i^{s+1} = 1 \qquad (4.13)$$

Flight t therefore represents the next flight that is not covered by one of rotations $x_1, x_2, \ldots, x_s$. Flight t must be covered by one of the rotations (columns) belonging to the t-th block. We assume that this flight is covered by some column before the t-th block. Let it be some column from the q-th block. The q-th block also covers the q-th flight that was covered earlier (before level s) which is not allowed. Therefore, it is clear that flight t must be covered by one of the columns from the t-th block. For all columns k located between the s-th column and the first column of the t-th block, we can say $\bar{x}_k=0$. We still must determine which is the first column of the t-th block. We denote the number of columns in block S_p by b_p (the number of columns in each block is very easy to establish upon formation of matrix $[a_{ij}]$). We denote the last column in block S_p by B_p. It is clear that :

$$B_p = \sum_{m=1}^{p} b_m \tag{4.14}$$

In accordance with expression (4.14), we conclude that the last column of the $(t-1)$-th block is B_{t-1}, and that the first column of the t-th block is $B_{t-1} + 1$.

We can conclude that if t is the smallest index of i for which $r_i^{s+1}=1$ after finishing our investigation on level s, then $\bar{x}_k=0$ for $s<k<B_{t-1}$ and the next column to be considered for the optimal solution is column $s'=B_{t-1}+1$. In addition, $\underline{R}_s = \underline{R}_{s+1}$. As can be seen, numerous columns are skipped over in this case which considerably speeds up the process of finding the optimal solution. Let us note some fight i for which $r_i^s = 1$ on level s. This means that flight i is not covered by any rotation before level s. Flight i can be covered by different rotations. The last rotation that can cover it is the last column in the i-th block, i.e. the column with index B_i. If :

$$B_i < s$$

we conclude that there is an i-th flight that must be covered before level B_i which is one of the levels before level s on which we are presently located. In this case, we leave level s and return to level $s-1$ and all paths that lead from the node on level s do not have to be considered further.

We denote by X^O the best feasible solution known to us at a specific moment in our search for the optimal solution. Solution X^O has the lowest costs of all feasible solutions that have been discovered. We denote these costs by F^O. Let us assume that we have reached level s . The costs of the rotations chosen from the first $s-1$ level equal :

$$\sum_{j=1}^{s-1} c_j \bar{x}_j$$

We can accelerate the search for the optimal solution by only investigating potential solutions whose costs are less than costs F^O . In other words, we are only interested in potential solutions for which :

$$\sum_{j=1}^{s-1} c_j \bar{x}_j + \sum_{j=s}^{n} c_j x_j < F^O \tag{4.15}$$

and :

$$\sum_{j=s}^{n} c_j x_j < F^O - \sum_{j=1}^{s-1} c_j \bar{x}_j \tag{4.16}$$

The solution shown to be better than the best feasible solution known so far then becomes the new best feasible solution.

Costs $(F^O - \sum_{j=1}^{s-1} c_j \bar{x}_j)$ denoted by f_s . If on some level :

$$c_s \geqslant f_s \tag{4.17}$$

we conclude that $\bar{x}_s = 0$, since including this rotation in the solution would lead to costs greater than costs F^O. In this case we also stop investigating all paths containing branch $\bar{x}_s=0$ and return back to level $s-1$.

We denote by t the smallest index i for which $r_i^{s+1}=1$. In other words, after passing through the first s levels, t is the next flight to be covered. As we noted and explained earlier, flight t must be covered by one of the columns from block t. The first column of block t is column $B_{t-1}+1$, and the last column is B_t. We denote by a_t the costs of some column from block t which are less than the costs of any other column in block t, i.e. :

$$a_t = \min_{B_{t-1}+1 \leqslant j \leqslant B_t} \{ c_j \} \tag{4.18}$$

If :

$$a_t \geqslant f_{s+1} \tag{4.19}$$

we conclude that in this case we should also discontinue any further search along this path.

Let us consider one more test that eliminates a large number of potential solutions from further appraisal and hastens the process even more. After passing through the first $s-1$ levels, let there be k_s flights left uncovered. Since we endeavor in our choice of rotation on the s-th, $s+1$-th,..., n-th level to choose a rotation that will cover the remaining k_s flights with the least possible costs, we must necessarily solve the following problem :

$$G_s = \sum_{j=1}^{n} c_j z_j \rightarrow min \qquad (4.20)$$

$$\sum_{j=1}^{n} h_j z_j = k_s \qquad (4.21)$$

$$z_j = 0 \quad or \quad 1 , \quad j = s, s+1, \ldots, n \qquad (4.22)$$

The test that eliminates a large number of potential solutions is comprised of the following. If :

$$G_s \geqslant f_s \qquad (4.23)$$

the search for the optimal solution is stopped along the path where we are located. In order to make the test (4.23), the problem defined by expressions (4.20), (4.21) and (4.22) must first be solved. This problem can be solved by combinatorial or dynamic programming. Often the lower boundary of value G_s is estimated instead of solving this problem. We denote this lower boundary by $\bar{G}_s$. We assume that we have in some way established the value of $\bar{G}_s$. Value G_s cannot take on a lesser value than $\bar{G}_s$ which is why $\bar{G}_s$ is called the lower boundary of G_s . The test in (4.23) can now be replaced by the following test. If :

$$\bar{G}_s \geqslant f_s \qquad (4.24)$$

we stop our search for the optimal solution along the path where we are located. A logical question is how to establish the value of $\bar{G}_s$. Pierce[112] proposed the following manner :

$$\bar{G}_s = m_s \cdot a_s \qquad (4.25)$$

where :

m_s - the number of uncovered flights on level s

$$a_s = \min_{j \geq s} \left\{ \frac{c_j}{h_j} \right\} \qquad (4.26)$$

Value a_s denotes the smallest possible costs per uncovered flight. Value $\bar{G}_s$ represents the smallest possible costs that can appear when going from level s to level n . Pierce and Lasky [114] used the second expression for $\bar{G}_s$, thereby shortening total computer run time.

The following numerical example illustrates the above procedure to find a set of rotations that cover all planned fights exactly once with the least possible costs :

Let there be 5 flights to cover and 31 possible rotations. Matrix $[a_{ij}]$ reads :

POSSIBLE ROTATIONS

FLIGHTS	1	2	3	4	5	6	7	8	9	10	11	12	13	14	15	16	17	18	19	20	21	22	23	24	25	26	27	28	29	30	31
1	1	1	1	1	1	1	1	1	1	1	1	1	1	1	1	1	0	0	0	0	0	0	0	0	0	0	0	0	0	0	0
2	1	1	1	1	0	1	1	0	1	0	0	1	0	0	0	0	1	1	1	1	1	1	1	1	0	0	0	0	0	0	0
3	1	1	1	0	1	1	0	1	0	1	0	0	1	0	0	0	1	1	1	0	1	0	0	0	1	1	1	1	0	0	0
4	1	1	0	1	1	0	1	1	0	0	1	0	0	1	0	0	1	1	0	1	0	1	0	0	1	1	0	0	1	1	0
5	1	0	1	1	1	0	0	0	1	1	1	0	0	0	1	0	1	0	1	1	0	0	1	0	1	0	1	0	1	0	1
	S_1																S_2								S_3				S_4		S_5
	FIRST BLOCK																SECOND BLOCK								THIRD BLOCK				FOURTH BLOCK		FIFTH BLOCK

The costs for individual rotations, the number of flights in the rotation, rotation costs per flight covered by the rotation, the last column in each block and the cost of the least expensive rotation of each block are given below.

TABLE 4.2

Rotation i	Rotation costs c_i	No. of flights in rotation h_i	Costs flight covered by rotation $\frac{c_i}{h_i}$	Block rotation belongs to S_p	Last column in block B_p	Costs of cheapest rotation in block a_p
1	66,5	5	13,3	1	16	25
2	54	4	13,5	1	16	25
3	58	4	14,5	1	16	25
4	57	4	14,25	1	16	25
5	56	4	14	1	16	25
6	48	3	16	1	16	25
7	57	3	19	1	16	25
8	45	3	15	1	16	25
9	48	3	16	1	16	25
10	54	3	18	1	16	25
11	57	3	19	1	16	25
12	44	2	22	1	16	25
13	40	2	20	1	16	25
14	38	2	19	1	16	25
15	46	2	23	1	16	25
16	23,5	1	23,5	1	16	25
17	40	4	10	2	24	18
18	36	3	12	2	24	18
19	40	3	13,33	2	24	18
20	39	3	13	2	24	18
21	30	2	15	2	24	18
22	34	2	17	2	24	18
23	32	2	16	2	24	18
24	18	1	18	2	24	18
25	63	3	21	3	28	27
26	48	2	24	3	28	27
27	50	2	25	3	28	27
28	27	1	27	3	28	27
29	70	2	35	4	30	40
30	40	1	40	4	30	40
31	32	1	32	5	31	32

The next step is to line up the rotations within each block by increasing order of cost per flight covered by the rotation. This is done in Table 4.3.

TABLE 4.3

Rotation		Rotation costs	No. of flights in rotation	Costs/flight covered by rotation	Block rotation belongs to	Last column in block	Costs of cheapest rotation in block
	Old No. rotation						
i		c_i	h_i	$\frac{c_i}{h_i}$	S_p	B_p	a_p
1	(1)	66,5	5	13,3	1	16	25
2	(2)	54	4	13,5	1	16	25
3	(5)	56	4	14	1	16	25
4	(4)	57	4	14,25	1	16	25
5	(3)	58	4	14,5	1	16	25
6	(8)	45	3	15	1	16	25
7	(6)	48	3	16	1	16	25
8	(9)	48	3	16	1	16	25
9	(10)	54	3	18	1	16	25
10	(7)	57	3	19	1	16	25
11	(11)	57	3	19	1	16	25
12	(14)	38	2	19	1	16	25
13	(13)	40	2	20	1	16	25
14	(12)	44	2	22	1	16	25
15	(15)	46	2	23	1	16	25
16	(16)	23,5	1	23,5	1	16	25
17	(17)	40	4	10	2	24	18
18	(18)	36	3	12	2	24	18
19	(20)	39	3	13	2	24	18
20	(19)	40	3	13,33	2	24	18
21	(21)	30	2	15	2	24	18
22	(23)	32	2	16	2	24	18
23	(22)	34	2	17	2	24	18
24	(24)	18	1	18	2	24	18
25	(25)	63	3	21	3	28	27
26	(26)	48	2	24	3	28	27
27	(27)	50	2	25	3	28	27
28	(28)	27	1	27	3	28	27
29	(29)	70	2	35	4	30	40
30	(30)	40	1	40	4	30	40
31	(31)	32	1	32	5	31	32

When lining up the rotations by increasing order of cost, there are changes in the order of some rotations. Thus, rotation 5 becomes rotation 3, rotation 8 becomes rotation 6, etc. The new matrix $[a_{ij}]$ is shown on the following page.

We denote by 0 the initial node from which branching starts. From node 0 we can go either to node 1 or node 2. Node 1 refers to the situation when the first rotation is chosen to be included in the solution ($x_1=1$), while node 2 denotes the situation when rotation 1 is not included in the solution ($x_1=0$). Nodes 0, 1 and 2 are shown on Figure 4.4.

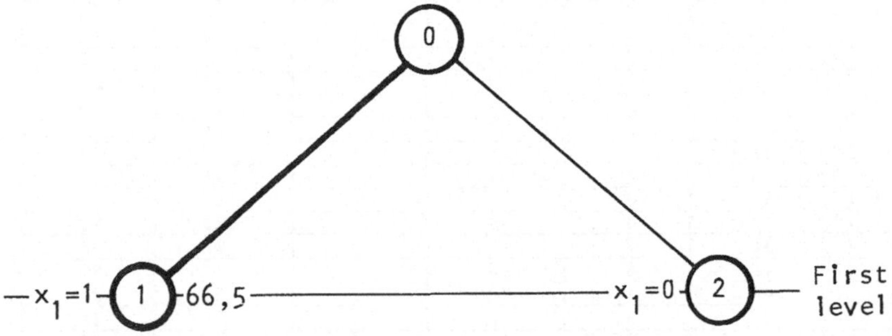

FIGURE 4.4 Nodes 0, 1 and 2.

Rotation 1 covers all 5 planned flights exactly once with costs equalling 66.5 (the cost of rotation 1). The solution $x_1=1$, $x_i=0$, $i=1,2,\ldots,31$ is feasible. This solution is also the best solution known to us thus far. The cost of the best-known feasible solution is :

$$F^0 = 66{,}5$$

POSSIBLE ROTATIONS

FLIGHTS	1	2	3	4	5	6	7	8	9	10	11	12	13	14	15	16	17	18	19	20	21	22	23	24	25	26	27	28	29	30	31
1	1	1	1	1	1	1	1	1	1	1	1	1	1	1	1	1	0	0	0	0	0	0	0	0	0	0	0	0	0	0	0
2	1	1	0	1	1	0	1	1	0	1	0	0	0	1	0	0	1	1	1	1	1	1	1	1	0	0	0	0	0	0	0
3	1	1	1	0	1	1	1	0	1	0	0	0	1	0	0	0	1	1	0	1	1	0	0	0	1	1	1	1	0	0	0
4	1	1	1	1	0	1	0	0	0	1	1	1	0	0	0	0	1	1	1	0	0	0	1	0	1	1	0	0	1	1	0
5	1	0	1	1	1	0	0	1	1	0	1	0	0	0	1	0	1	0	1	1	0	1	0	0	1	0	1	0	1	0	1

The values of r_i^2 equal :

$$r_i^2 = 0, \qquad i = 1, 2, 3, 4, 5$$

We return to the initial state (node 0) and drop down to the first level in node 2 ($x_1=0$).

Total costs in node 2 are zero. We drop down to the second level in nodes 3 and 4 (Figure 4.5).

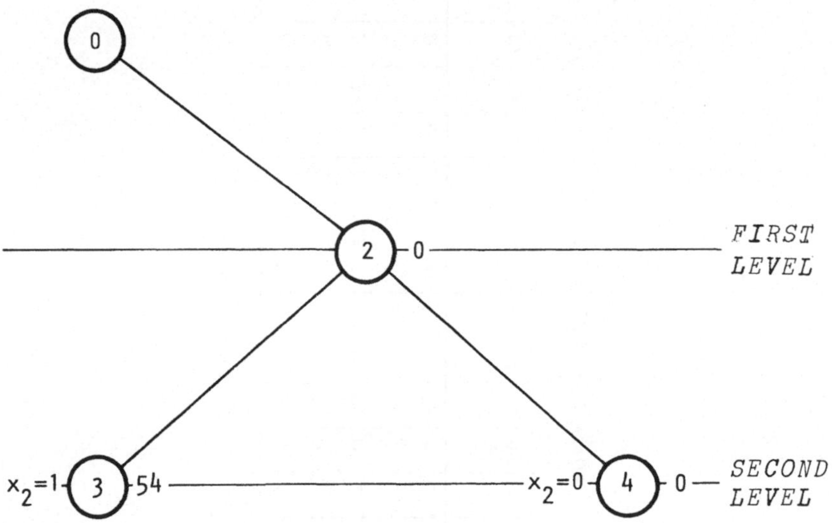

FIGURE 4.5 Nodes 0, 2, 3 and 4

Total costs in node 3 are 54 (the costs of rotation 2). Values r_i^3 $(i=1,2,3,4,5)$ equal :

$$r_1^3 = 0, \ r_2^3 = 0, \ r_3^3 = 0, \ r_4^3 = 0, \ r_5^3 = 1$$

since rotation 2 covers all flights except flight 5.

Considering that $r_5^3=1$, flight 5 must be covered by some other rotation from the fifth block. Since the fifth block only contains rotation 31, the fifth flight must be covered by rotation 31. This means that $x_i=0$ for $i=3,4,\ldots,30$. Thus all flights are covered exactly once. However, because :

$$f_s = F^0 - \sum_{j=1}^{s-1} c_j \bar{x}_j ,$$

i.e.

$$f_{31} = 66,5 - \sum_{j=1}^{30} c_j \bar{x}_j = 66,5 - 54 = 12,5$$

and

$$c_{31} = 32 > 12,5 = f_{31}$$

we conclude that we cannot get a better solution than the one we already have. The total costs of solution $x_1=0, x_2=1, x_i=0$ for $i = 3,4,\ldots,30$ that we are rejecting would be 54+32=86.

We return to node 2 from node 3 and drop down from node 2 to node 4, whence we further branch to nodes 5 and 6 on level 3 (Figure 4.6).

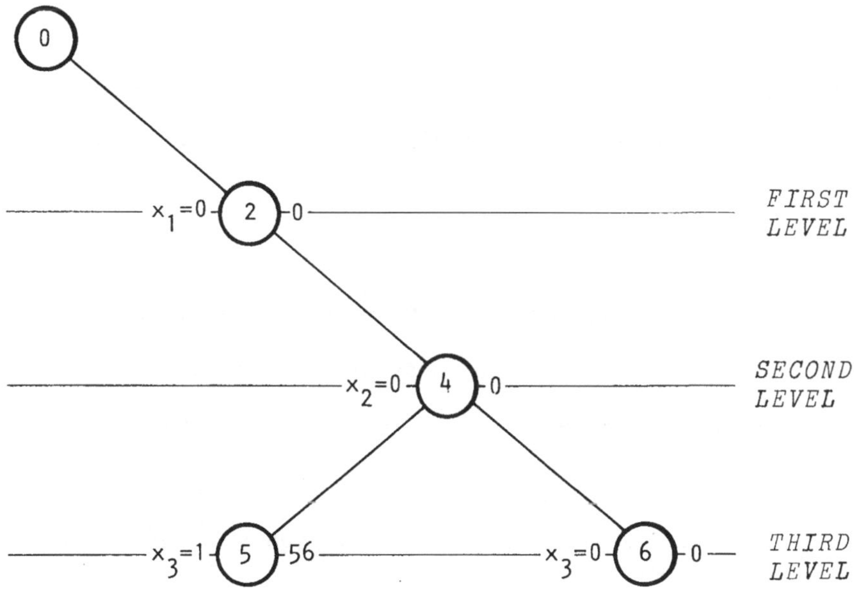

FIGURE 4.6 Nodes 0, 2, 4, 5 and 6.

Values r_i^4 $(i=1,2,3,4,5)$ equal :

$r_1^4 = 0,\ r_2^4 = 1,\ r_3^4 = 0,\ r_4^4 = 0,\ r_5^4 = 0$

since rotation 3 covers all flights except flight 2. Considering that $r_2^4=1$, we conclude that flight 2 must be covered by a rotation from the second block. The second block starts with rotation number 17. Therefore, $x_i=0$ for $i=4,5,6,$ $,\ldots,16$. Within the second block, the only rotation that can be taken into consideration is 24. Rotations 17, 18,...,23 cannot be considered since choosing any of these rotations would result in some flights being covered twice. Rotation 24 only covers flight 2, which means that all flights are covered exactly once with rotations 3 and 24. This is :

$$f_{24} = 66,5 - \sum_{j=1}^{23} c_j \bar{x}_j = 66,5 - c_3 = 66,5-56 = 10,5$$

$$c_{24} = 18 > 10,5 = f_{24}$$

We conclude that we cannot obtain a better solution than the one we already have and so we reject solution $x_1=0, x_2=0, x_3=1$ $x_i=0$ for $i=3,4,\ldots,23, x_{24}=1, x_i=0$ for $i=25,26,\ldots,31$.

By applying the above procedure to further branching from node 6, we see it is not possible to obtain a feasible solution that would entail costs less than 66.5 for the situation when $x_4=1,\ x_5=1,\ x_6=1,\ x_7=1,\ x_8=1,\ x_9=1$ $x_{10}=1, x_{11}=1, x_{12}=1, x_{13}=1, x_{14}=1, x_{15}=1$. Figure 4.7 presents the nodes we have examined up to the sixteenth level.

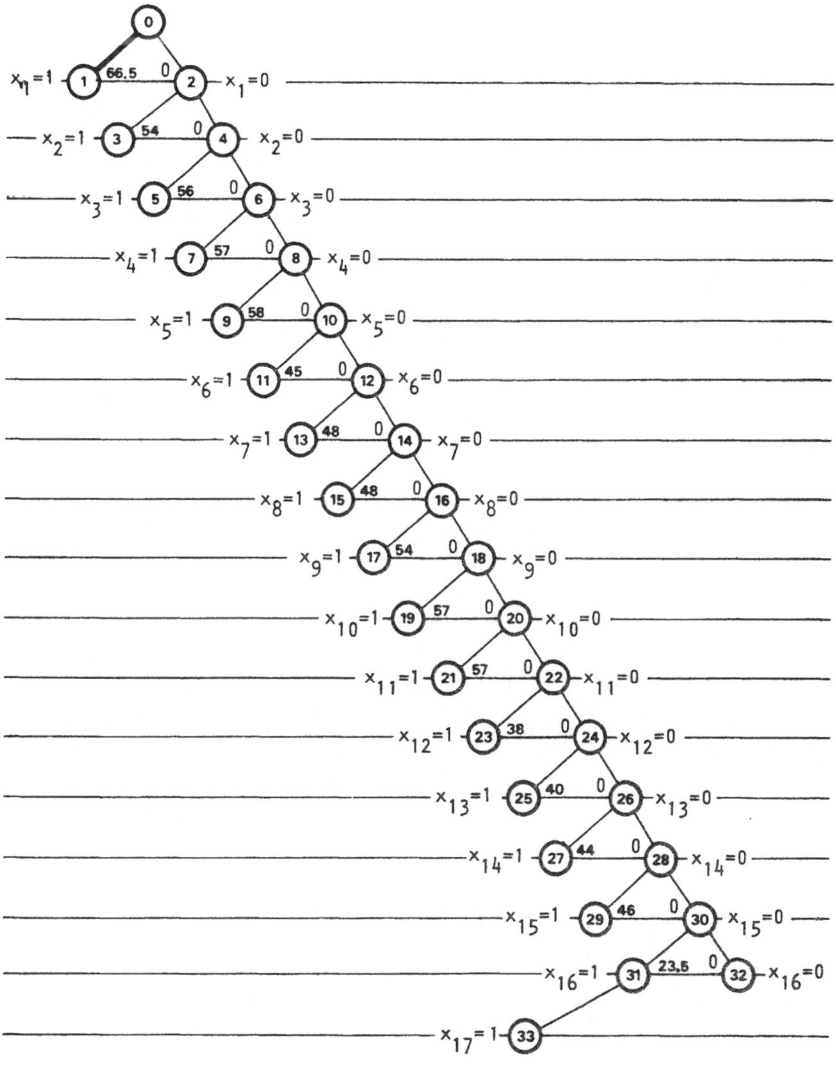

FIGURE 4.7 Nodes investigated to the 16th level.

The sixteenth level contains nodes 31 and 32. Node 31 concerns the situation when $x_{16}=1$. Rotation 16 only covers flight no. 1. This means that :

$$r_1^{17} = 0, \quad r_2^{17} = 1, \ r_3^{17} = 1, \ r_4^{17} = 1, \ r_5^{17} = 1$$

The smallest i for which $x_i^{17}=1$ is 2. Flight 2 is therefore the next flight that is not covered. This flight must be covered by some rotation from the second block. The first rotation in the second block is rotation 17, which covers the second, third, fourth and fifth flights, which means that rotations 16 and 17 cover all flights exactly once. Thus :

$$f_{17} = 66,5 - \sum_{j=1}^{16} c_j \bar{x}_j = 66,5-c_{16}=66,5-23,5 = 43$$

$$c_{17} = 40 < 43 = f_{17}$$

which means that the path containing node 31 ($x_{16} = 1$) provides a solution that is better than the best feasible solution known. Total costs of rotations 16 and 17 are $c_{16}+c_{17}=23,5+40=63,5$. Since :

$$c_{16} + c_{17} = 63,5 < 66,5 = c_1 = F^o$$

solution $x_i=0$, $i=1,2,\ldots,15$, $x_{16}=1$, $x_{17}=1$, $x_i=0$, $i=18,19,\ldots,31$ becomes the new best feasible solution discovered thus far.

We will not examine the other rotations from the second block as only rotation 17 from the second block can be matched with rotation 16 so that all flights are covered only once. This means that if $x_{16}=1$, then $x_{17}=1$. This excludes from any further consideration all paths containing the pair of nodes $x_{16} = 1$, $x_{17}= 0$. We return to the fifteenth level and drop down to node 32 on the sixteenth level in order to continue branching.

By applying this procedure to further branching through the 31st level, no solution is found with smaller costs than 63.5 which means that solution $x_i=0$, $i=1,2,\ldots,15$, $x_{16}=1, x_{17}=1$ $x_i=0$, $i=18,19,\ldots,31$ is the optimal solution. Figure 4.8 shows all nodes examined until the optimal solution was found.

The tree that would be obtained if all theoretically possible solutions were generated would contain $2^{32}- 1$ nodes. The efficiency of the procedure presented lies in the fact that of the theoretically possible $2^{32} - 1$ nodes, only 63 nodes were examined until the optimal solution was found.

Pierce and Lasky [114] improved the Pierce model. Let us take a brief look at the improvement to the Pierce [112] model described above. Pierce and Lasky proposed that blocks $S_1, S_2, \ldots, S_m$ be organized in the following manner.

First, all n possible rotations must be lined up in increasing order of costs per flight covered by the rotation. In other words, the rotations should be lined up so that :

$$\frac{c_1}{h_1} \leqslant \frac{c_2}{h_2} \leqslant \ldots \leqslant \frac{c_n}{h_n} \tag{4.27}$$

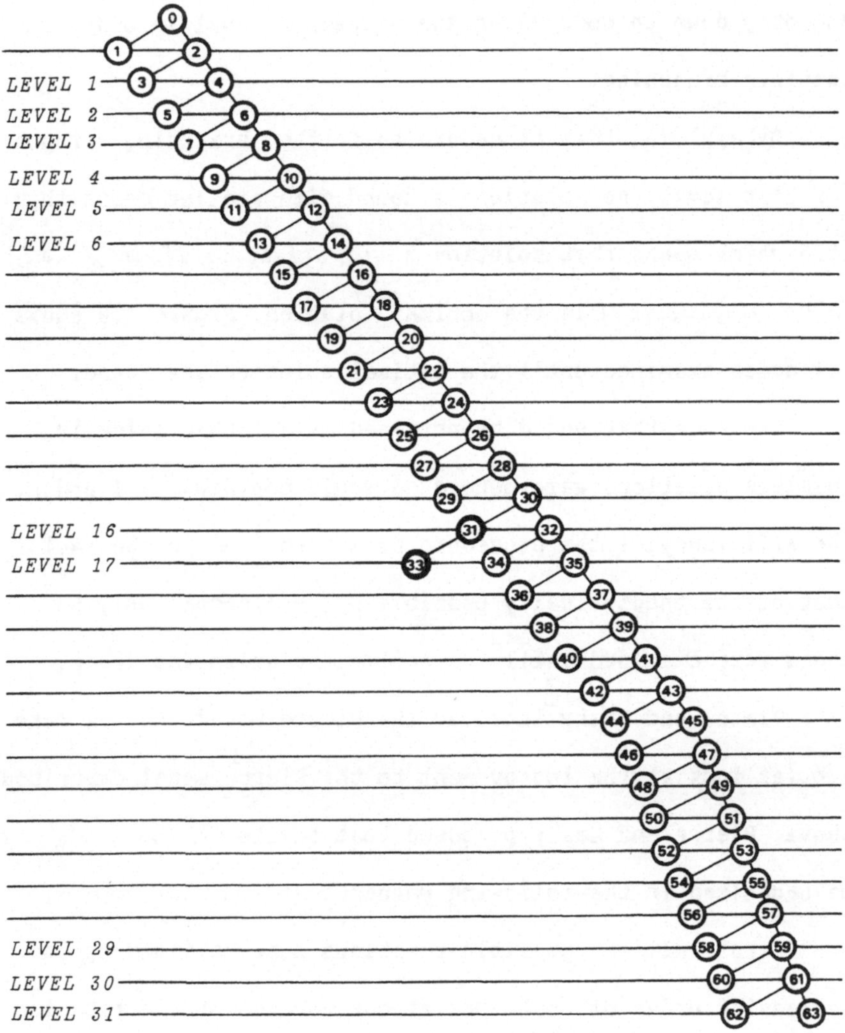

FIGURE 4.8 **Nodes examined until the optimal solution was found**

We include rotation A_1 corresponding to the lowest costs per flight covered by the rotation to block S_1. We also include rotation A_2 to block S_1 if there is at least one flight that is simultaneously covered by rotation A_1 and rotation A_2. If there is no such flight that is simultaneously covered by rotation A_1 and rotation A_2, then rotation A_2 should be included in block S_2. We first assume that both rotation A_1 and rotation A_2 are in block S_1. We will also include rotation A_3 in block S_1 if there is at least one flight that is covered by rotation A_1, rotation A_2 and rotation A_3. If there is no such flight then rotation A_3 is included in block S_2. When rotation A_1 is in block S_1, and rotation A_2 in block S_2, then we must first examine whether there is at least one flight simultaneously covered by rotation A_1 and rotation A_3. Should such a flight exist, we include rotation A_3 in block S_1. If such a flight does not exist, we further examine whether there is a flight covered simultaneously by rotation A_2 and rotation A_3. If such a flight exists, rotation A_3 is included in block S_2, and if not then rotation A_3 is included in block S_3. Thus, every rotation joins the first block whose column with its rotation covers at least one common flight. This manner of forming blocks was shown to be more efficient (in terms of computer run time) than the manner proposed by Pierce. To this effect, block order in the

Pierce model depended on the flight order. The block that is first, second, etc., essentially depends on which flight is first, second, etc. Since the flights are in arbitrary order, we conclude that the blocks are also in arbitrary order. The Pierce and Lasky modification concerning how to form the blocks decreases the computer time needed to solve the crew scheduling problem. Pierce and Lasky were also able to decrease the computer run time by making a better estimation of value $\bar{G}_s$. In expression (4.25), value $\bar{G}_s$ is defined as $\bar{G}_s = m_s * a_s$. Pierce and Lasky proposed that value $\bar{G}_s$ be calculated as follows : We denote by e_p the maximum number of flights that can be covered by one of the columns from set S_p:

$$e_p = \max_{j \in S_p} \{h_j\} \tag{4.28}$$

We denote by d_p the least possible costs of covering a flight by one of the columns from set S_p. This means that :

$$d_p = \min_{j \in S_p} \left\{ \frac{c_j}{h_j} \right\} \tag{4.29}$$

The least possible costs W_p when covering n_p flights equals:

$$W_p = e_p \cdot d_p \tag{4.30}$$

Let $i_1, i_2, \ldots, i_q$ be indexes of flights that are not yet covered at the moment we are located on level s (the values of $\bar{x}_1, \bar{x}_2, \ldots, \bar{x}_{s-1}$ taken by the first $s-1$ variables are already determined). We introduce values $\bar{e}_{(i_j)}$ and $\bar{w}_{(i_j)}$ into the discussion, defined as :

$$\bar{e}_{(i_j)} = \sum_{k=1}^{j} e_{(i_k)} \qquad , \; j = 1,2,\ldots,q \qquad (4.31)$$

$$\bar{W}_{(i_j)} = \sum_{k=1}^{j} W_{(i_k)} \qquad , \; j = 1,2,\ldots,q \qquad (4.32)$$

Values $\bar{e}_{(i_j)}$ and $\bar{w}_{(i_j)}$ are defined by the function shown in Figure 4.9.

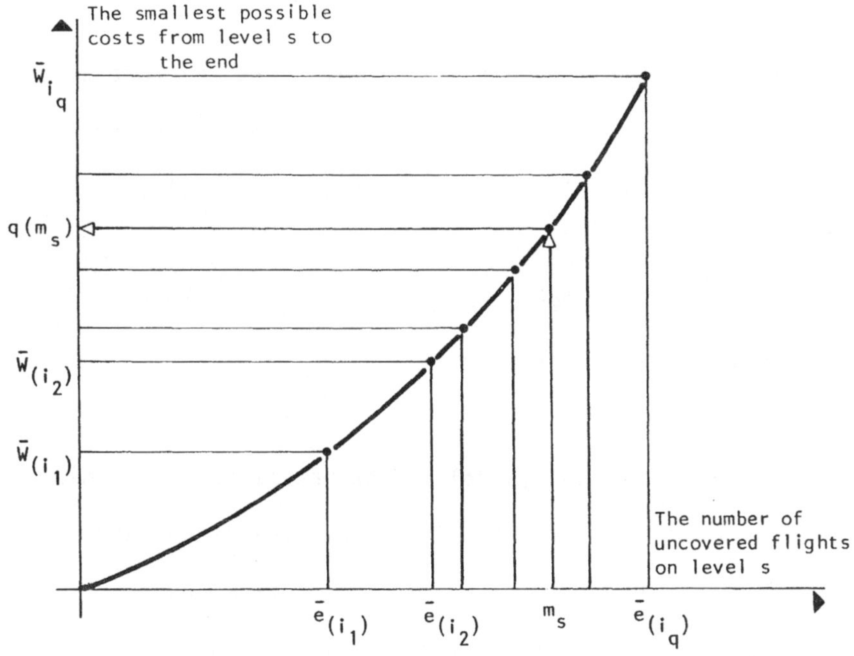

FIGURE 4.9 Minimum costs from level s to the end as a function of uncovered flights on level s

Value $\bar{w}_{(i_j)}$ represents the smallest possible cost that can be expected from level s to the end in order to cover the remaining $\bar{e}_{i_j}$ flights. For a known number m_s of uncovered flights on some level s, the function shown on Figure 4.9 enables us to estimate the smallest possible costs $q(m_s)$ we can have from level s to the end in order to cover the uncovered m_s flights. Better computer run time can be achieved by using value $q(m_s)$ instead of value $\bar{G}_s$.

We denote by k_s total costs concluding with level $s-1$:

$$k_s = \sum_{j=1}^{s-1} c_j \, x_j \tag{4.33}$$

Pierce and Lasky [114] proposed the following test in order to accelerate the process of finding the optimal solution. If we have arrived at level s (we have determined values $\bar{x}_1, \bar{x}_2, ., \bar{x}_{s-1}$), then if :

$$k_s + q(m_s) \geq F^o \tag{4.34}$$

we stop searching for the optimal solution along the path where we are located.

Should expression (4.34) not be satisfied, we continue our search for the optimal solution by examining which of rotations A_j from set S_{i_1} (i_1 is the index of the next uncovered flight) should be included in the solution. The first by rotation order in this block is rotation A_{i_1+1} . The rotation costs per flight covered ($\dfrac{c_{i_1+1}}{h_{i_1+1}}$) are the least of all costs per covered flight by a rotation belonging to set S_{i_1}. (Rotations are lined up by increasing order of costs). The first rotation A_j from set S_{i_1} that does not satisfy expression (4.34) and that satisfies the following expression :

$$k_s + c_j + (m_s - h_j)\left(\frac{c_{i_1+1}}{h_{i_1+1}}\right) < F^o \qquad (4.35)$$

is included in the solution.

Should rotation A_j not satisfy expression (4.35) and if:

$$k_s + \bar{c}_j \geqslant F^o \quad , \quad \bar{c}_j = \min_{k \in S_{i_1};\ k > j} \{c_k\} \qquad (4.36)$$

there is no further examination of rotations from set S_{i_1} since even including the rotation with the lowest costs per flight would result in a solution with greater costs than the costs of the best known feasible solution.

Should neither expression (4.35) nor (4.36) be satisfied, investigation continues of rotation A_j from set S_{i_1}, since there is a chance of finding a solution with lower costs than those of the best known feasible solution.

The Marsten[97] model that also solves the crew scheduling problem is similarly based on the branch-and-bound method. Marsten et al[98] and Marsten and Shepardson[99] presented the experience of implementing crew scheduling models in Flying Tiger Lines, Pacific Southwest Airways and Continental

Airlines. These models are based on Marsten's results. As noted by Marsten et al[98] and Marsten and Shepardson,[99] treating the crew scheduling problem as a set partitioning problem can only give satisfactory results when the problem has smaller dimensions. The problems mentioned by Marsten have, for example, 400 flights and 30,000 possible rotations and cannot be solved as a set partitioning problem. Therefore, the methodology presented in this chapter can be applied for smaller air carriers, for smaller organizational units of larger air carriers or for larger air carriers where the rotations are of very short duration. Various heuristic methods are most often used to solve crew scheduling problems for larger air carriers. Ball and Roberts[11] recently developed a model whereby the crew scheduling problem is treated as a graph partitioning problem. This model, which will be considered later on, has shown to be very efficient from the computer point of view.

4.5. Heuristic Set Covering Algorithms for Large Airline Crew Scheduling Problems

Set partitioning and set covering problems belong to the category of NP hard problems. For these problems, computer time rises exponentially, not polynomially, with an increase in the problem's dimensions. As was mentioned earlier, treating the crew scheduling problem as a set partitioning or set covering problem can only provide satisfactory results for smaller air carriers, smaller organizational units of larger air carriers or for larger air carriers whose rotations are of very short duration. The main reason why set partitioning and set covering problems cannot solve crew scheduling problems for larger air carriers is that these are NP hard problems and the optimal solution for larger air carriers cannot be obtained within an acceptable period of time. Different heuristic algorithms can be used to solve these problems for larger air carriers. Input data for the heuristic procedure comprise planned flights, i.e. the airline schedule. The next step is to generate possible rotations in which existing legal and company constraints are taken into consideration. For example, frequent operational constraints include maximum allowed time away from the base (rotations where this time is greater than a set amount are no longer considered), or time away from the base and working time during that period, etc.

After generating the possible rotations, a heuristic set covering algorithm is applied to obtain the solution, i.e. the crew schedule. The next step is to try a large number of changes to the solution in order to get a better and better solution, i.e. one with lower and lower costs. The final crew schedule should be adopted when the possibilities have been exhausted of obtaining a solution with lower costs. The quality of the solution obtained (in terms of the amount of costs it entails) and the speed at which the first solution is obtained, certainly depend on the heuristic algorithm applied. Baker et al[9] recently developed 4 different set covering heuristic algorithms. These authors treated the crew scheduling problem as a set covering problem, i.e. when some flights can have crew members sitting in passenger seats going to or returning from duty. The 4 different heuristic algorithms were denoted respectively by letters A, B, C, D .

At the beginning of algorithm A , the set of chosen rotations that will enter into the final solution does not contain any rotations. This set is gradually joined by different rotations thereby minimizing the costs per un-covered flight (an uncovered flight is one that has not been included in a rotation that is part of the final solution).

When algorithm *B* is applied, all chosen rotations are first lined up in increasing order of costs per flight covered. Rotations are added one by one to the set of rotations that will participate in the final solution, bearing in mind that no rotation can be added that contains a flight that has already been covered. At the end, when it is no longer possible to obtain a solution in this manner, the process is repeated to the effect that it is now permitted for a rotation to contain at most one earlier covered leg. If this also does not produce a solution, the procedure is repeated and two earlier covered legs are allowed, etc.

At the beginning of algorithm *C* there are no rotations in the set of chosen rotations entering in the final solution. Possible chosen rotations are added one by one to the set of final rotations, bearing in mind the number of uncovered legs.

For algorithm *D* , the set of final rotations that is empty at the beginning is added to gradually by rotations that are chosen so as to correspond to the smallest ratio of covered to uncovered legs.

Computer experiments carried out by Baker et al [9] indicate that algorithm *A* provides a solution closest to the optimum (whereby the costs of these solutions are closest to the costs of the optimal solution). On the other hand, algorithm *B* was shown to be the most computer efficient, i.e. a solution is found the fastest by using algorithm B . It was also shown that attempts to improve solutions obtained by one of the above algorithms did not give an essentially better solution (improvements were often under 1%), and required at the same time considerable computer time. The Baker et al[9] model (algorithm A) was tested by the Federal Express Corporation. The results obtained demonstrate the exceptional efficiency of this heuristic procedure. For example, for the case of one base and 3,267 legs to be flown, a total of 15,410 possible rotations was generated in only 1 min. 30 sec, with algorithm A providing a solution in only 1 min. 19 sec. When applying the set covering heuristic algorithm proposed by Baker et al,[9] solutions near to optimal are also obtained in a very short time for problems of large dimensions.

4.6. Crew Scheduling as a Graph Partitioning Problem

Ball and Roberts[11] recently developed a model that solves the crew scheduling problem as a graph partitioning problem. The model is based on the authors' results dealing with the problem of planning drivers' schedules in public transportation.

When solving crew scheduling as a set partitioning problem, a very large number of possible rotations is first generated, and then a specific number of rotations are chosen to be made by the crew members. Although, as already mentioned, various methods are applied to reduce the number of rotations generated, very large computer time is concerned when the problem has larger dimensions.

Ball and Roberts[11] represented the planned legs in the form of nodes on a graph. Each leg represents one node (Figure 4.9a).

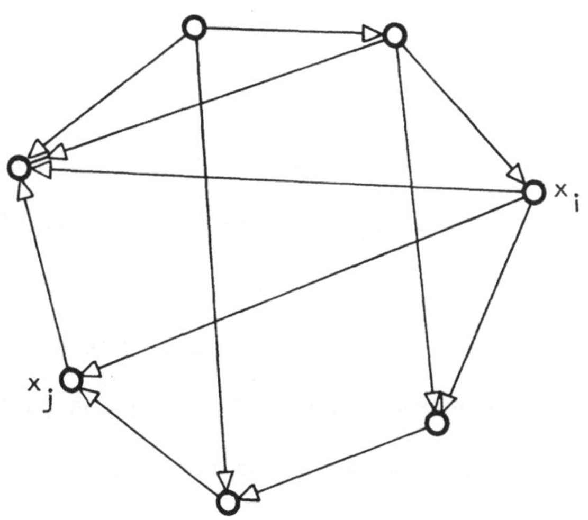

FIGURE 4.9a Graph $G(N,A)$ whose set of nodes comprise legs to be made

If graph $G(N,A)$ contains branches oriented from some node x_i to some node x_j, this means that the crew can fly leg x_j after completing leg x_i to the effect that no constraints are infringed upon regarding the crew's working time. As mentioned by Ball and Roberts, the main difference between graph partitioning and set partitioning problems lies in the fact that the size of the graph is proportional to the number of nodes or the number of nodes square, while the number of generated rotations in a set participation problem is most often an exponential function of the number of legs to be flown.

It is clear that several planned legs following one after the other in graph *G(N,A)* represent a path. Further discussion will only consider those rotations or series of legs flown by individual crews that do not breach regulated constraints linked to the crew's working time. Graph partitioning comprises finding the set of paths in the graph so that each node belongs to only one path. Since each path in graph *G(N,A)*(each rotation) is linked to certain costs, the crew scheduling problem formulated as a graph partitioning problem is composed of finding the set of paths in the graph whose total rotation costs are minimal, with each planned leg belonging to one path and no constraints being breached regarding crew working time. The crew scheduling problem formulated in this way means that each leg belongs to only one rotation, and that there is only one crew in the aircraft on each leg, the crew that is operating the aircraft. In reality, some crew members going to or returning from duty are often in the aircraft in passenger seats, so that the model must also include this type of situation. This will be discussed in more detail later on.

Let us first define the concept of matching in the graph. For given nonoriented graph $G(N,A)$, matching M represents a subset of the set of branches A, so that no two branches have a common node. If each node $i \in N$ is associated with costs a_i, and each branch $(i,j) \in A$ costs b_{ij} and if $N(M)$ denotes the set of nodes covered by matching M, then the problem of matching with least total costs can be formulated as :

$$\min \left\{ \sum_{(i,j) \in M} b_{ij} + \sum_{i \in N-N(M)} a_i \ ; \quad M \quad \text{is matching} \right\}$$

Let us note graph $G(N,A)$ shown on Figure 4.10.

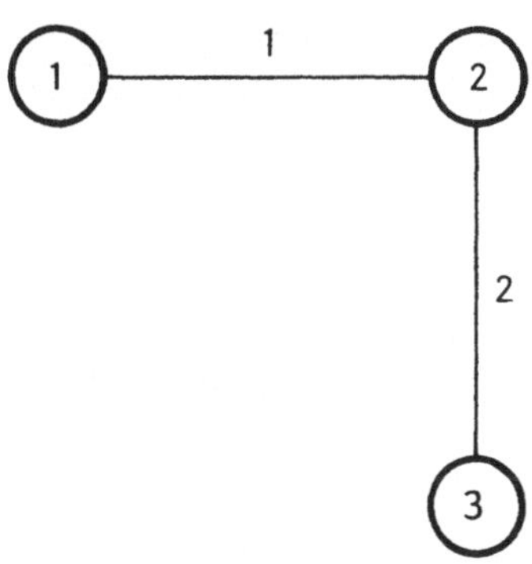

FIGURE 4.10 Graph illustrating the matching problem with minimum total costs

The graph shown on Figure 4.10 has three nodes denoted respectively by 1, 2, 3, and 2 branches denoted respectively by 1 and 2. The costs associated with individual nodes and branches are shown on Table 4.4.

TABLE 4.4

Node	*Costs*	*Branch*	*Costs*
1	5	1	2
2	3	2	6
3	4		

The problem of matching with the minimum total costs can be formulated mathematically as :

$$F = 2x_1+6x_2+5y_1+3y_2+4y_3 \rightarrow min \tag{4.37}$$

$$x_1 + y_1 = 1 \tag{4.38}$$

$$x_1 + x_2 + y_2 = 1 \tag{4.39}$$

$$x_2 + y_3 = 1 \tag{4.40}$$

Variables x_1, x_2 denoting branches and variables y_1, y_2, y_3 denoting nodes are binary variables that take on the value of 0 or 1. When, for example, $x_1 = 1$, branch 1 is in the matching with the least total costs. Also, when $y_3=1$ we conclude that node 3 exists as an isolated node in the matching with the least total costs. In addition, when $x_1=0$ we conclude that branch 1 does not exist in the matching with the smallest total costs. Solving the problems defined by expressions (4.37) - (4.40) produces the matching with the least total costs corresponding to the graph shown on Figure 4.10. The matchings with the least costs for this example are shown on Figure 4.11.

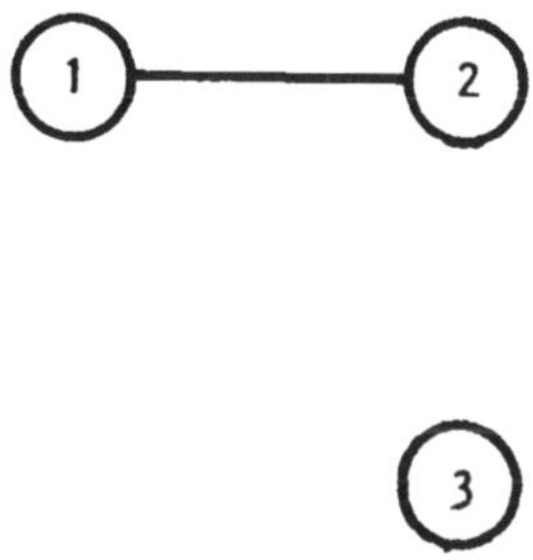

FIGURE 4.11 Matching with minimum total costs

The total cost of this matching is 6 (the costs of node 3 are 4, and the costs of branch 1 are 2). Matching with minimum total cost is one of the most important algorithmic steps in the Ball and Roberts algorithm that is used to solve the crew scheduling problem. In this algorithm, parts of individual rotations are first constructed in the following manner. First, individual nodes (legs to be flown) are grouped into levels. The node level is the length of the longest path leading to that node. We assume that the length of each branch in the graph equals 1 and denote by s the source node and by t the destination node. Nodes s and t are not planned legs to be flown (Figure 4.12).

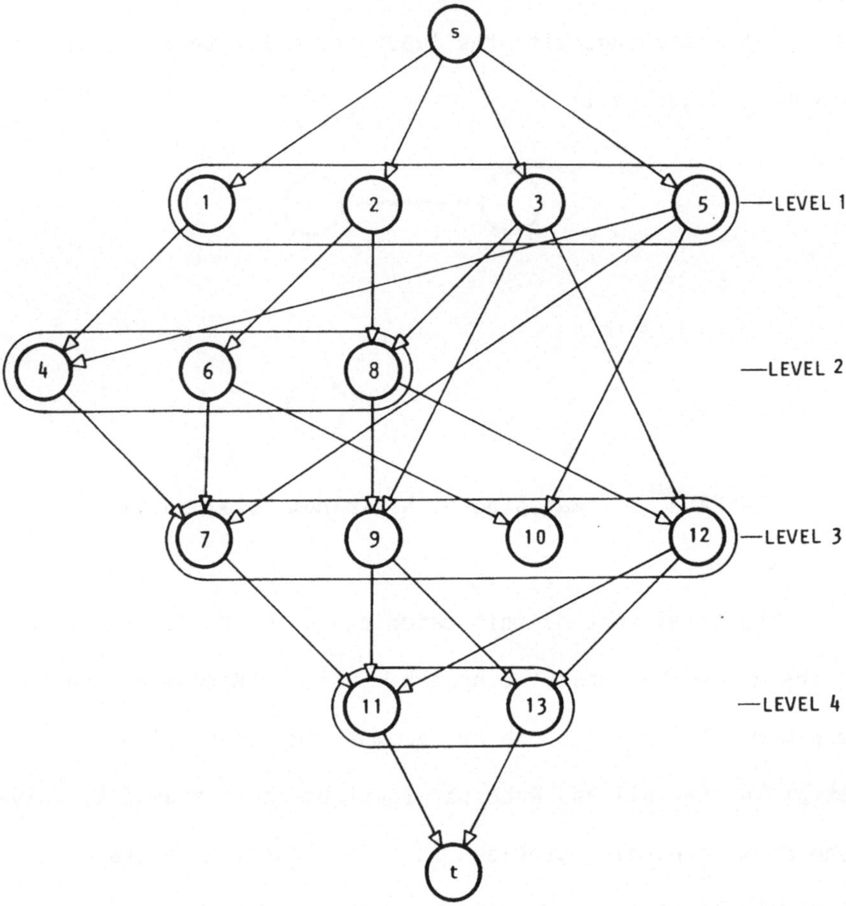

FIGURE 4.12 Grouping nodes into levels

Figure 4.12 shows source node s and destination node t and 12 planned legs to be flown. The first level contains nodes 1,2,3 and 5 since the length of the longest path from node s to them equals 1. Node 12, for example, belongs to level 3 since the longest path leading to it (path s, 2, 8, 12) equals 3. Figure 4.12 presents branches that can be used to couple nodes between individual levels. For example, node 2 on level 1 can be coupled with node 8 on level 2. It is also possible to couple node 3 on level 1 with node 8 on level 2. It is clear that node 8 on level 2 can be preceded by either node 2 or node 3 on level 1. The optimal matching of nodes on level 1 with nodes on level 2 (in the general case nodes on some level k with nodes on some level $k+1$) is found by solving the problem of matching with minimum total costs. For example, let the optimal matching be coupling node 1 with node 4, node 2 with node 6 and node 3 with node 8 (Figure 4.13).

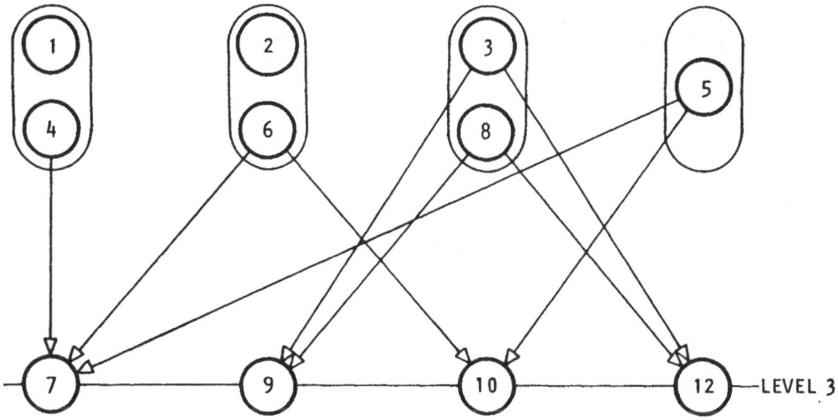

FIGURE 4.13 Optimal matching of nodes on the first and second levels

This has given us 4 rotations. The first rotation comprises for now leg 1 and leg 4, the second legs 2 an 6, etc. These rotation parts are treated hereinafter as nodes. The next step is to optimally pair these nodes with the nodes on level 3. It is important to note that when solving the problem of matching with least total costs, consideration is not given to so-called internal branches or branches that would lead to unfeasible solutions. For example, consideration is not given to branches leading from node 3 to node 9 and from node 3 to node 12 (Figure 4.13). These are internal branches, i.e. it was established in the previous iteration that node 3 precedes node 8, so there is no reason to consider the possibility of node 3 preceding node 9 or node 12. Also, if flying leg 10 after leg 6 breaches the constraint concerning crew working time, the branch leading form node 6 to node 10 is deleted from further consideration. In the next step, new nodes would be matched with the nodes on level 4, etc. As we can see, the construction of individual crew rotations is based on solving the problem of matching with the least total costs. Input data to find the optimal matching are the costs of individual branches and nodes. Every node or branch we consider represents a path. The costs of some path S are denoted by $d(S)$, the costs of node i are denoted by a_i, and the costs of branch (i,j) by b_{ij}, so that :

$a_i = d(S(i))$ - the costs of node i taking part in the matching,

$b_{ij} = d(S(i,j))$ - the costs of branch (i,j) taking part in the matching.

Let the costs of all nodes and all branches equal 1. In order to obtain the matching with the least total costs, as many nodes as possible must be joined into a partial rotation that also represents a node with costs equalling 1. Thus an attempt is made to minimize the number of rotations that are matched, i.e. to minimize the number of crews needed to carry out planned flights. However, in addition to minimizing the number of rotations, consideration must be given to the costs of the rotations and the time that the crew spends on the ground between two flights. We denote the costs of rotation S by $TC(S)$, and the total time the crew spends on the ground when carrying out rotation S by $GT(S)$. Ball and Roberts introduced the following variables into the discussion :

$$\bar{T} = \sum_{i \in N} TC\ (S(i)) + \sum_{(i,j) \in A} TC\ (S(i,j)) \qquad (4.41)$$

$$\bar{G} = \sum_{i \in N} GT\ (S(i)) + \sum_{(i,j) \in A} GT\ (S(i,j)) \qquad (4.42)$$

By defining factors a and b so that $a,b \geqslant 0$ and $a+b = 1$ the objective function is defined corresponding to some path S as :

$$1 + a \frac{TC(S)}{\bar{T}} + b \frac{GT(S)}{\bar{G}} \quad (4.43)$$

Thus, when solving the problem of matching with least total costs in each iteration, the least possible number of rotations (paths) is generated, to the effect that should there be several different solutions with an equal number of rotations, the one corresponding to the least value of the following expression is chosen :

$$a \frac{TC(S)}{\bar{T}} + b \frac{GT(S)}{\bar{G}} \quad (4.44)$$

We would also note that this expression can never be greater than 1. Ball and Roberts expanded the objective function defined by expression (4.43) by factor α that gives priority in matching to rotations S that start and finish in the same base.

The final form of the objective function associated with some rotation (path) S is :

$$1 + a \frac{TC(S)}{\bar{T}} + b \frac{GT(S)}{\bar{G}} + \alpha(S) \quad (4.45)$$

$$\alpha(S) = \begin{cases} \alpha', & \text{for which } \alpha'<0 \text{ if } S \text{ starts and finishes at the same base} \\ 0, & \text{otherwise} \end{cases} \quad (4.46)$$

This gives priority to rotations that start and finish in the same base because this reduces the number of crews that are sitting in passenger seats on their way to or from work duties.

Upon receipt of the set of rotations with the minimum objective function value, Ball and Roberts proposed a further possible decrease in the objective function value by making changes in the legs between individual rotations or by connecting several smaller rotations (Figure 4.14).

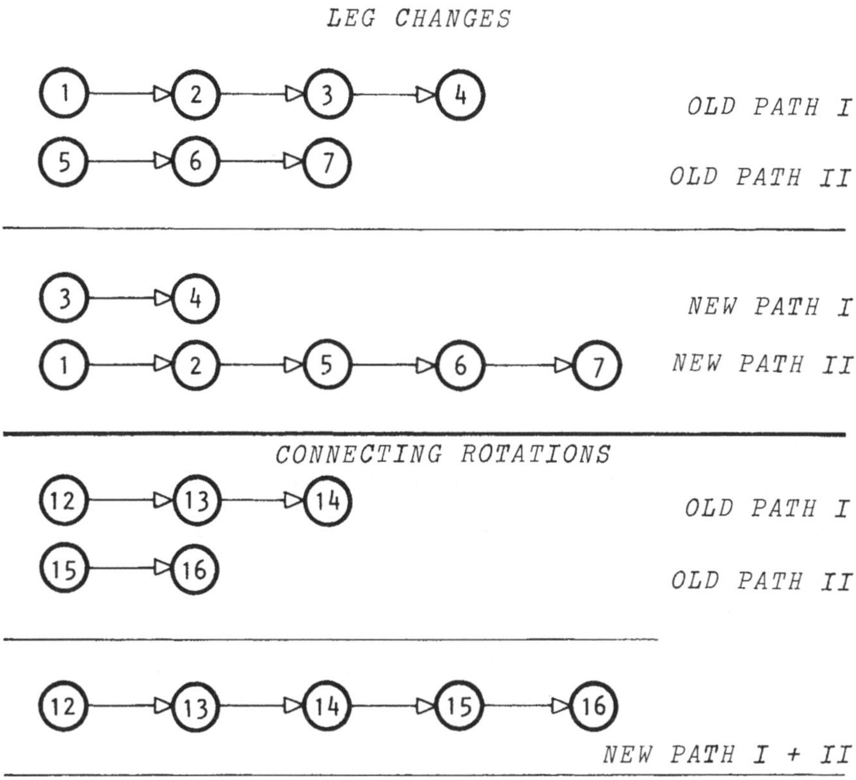

FIGURE 4.14 Changing legs and connecting rotations

Ball and Roberts also used the algorithm for finding the least total matching costs to find the parts of rotations to be optimally matched with parts of other rotations. In other words, whether or not a rotation should remain whole or whether certain of its parts should be connected with parts of other rotations is discovered through application of the algorithm for finding the matching with least total costs (Figure 4.15).

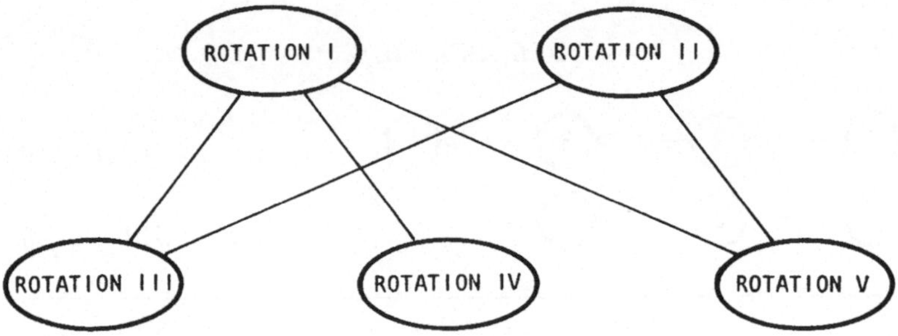

FIGURE 4.15 Possible rotation matchings

Figure 4.16 presents the solution obtained by applying the algorithm to find matchings with least total costs.

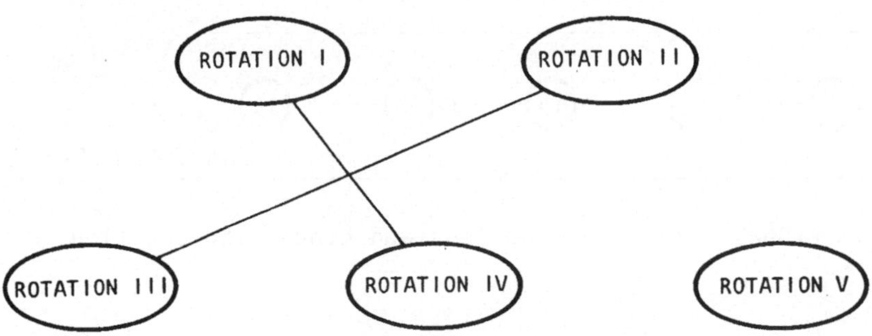

FIGURE 4.16 Optimal rotation matching

In our hypothetical example, the optimal solution would be to match part of rotation I with part of rotation IV, and part of rotation II with part of rotation III. Rotation V should remain unchanged. Input data for this phase comprise the costs of the rotations formed in the previous phase and the costs of matching parts of individual rotations in the new rotation.

In the final phase, Ball and Roberts considered the possibility of crews sitting in passenger seats when going to or coming from duty. Each formed rotation that does not start and finish in one of the air carrier's bases is given a leg at the beginning and/or end of the rotation in which a crew flying a formed rotation is located in the passenger seat (Figure 4.17).

The computer results Ball and Roberts obtained after testing indicate the efficiency of this model. In the case of a fleet of B-747 airplanes with 328 legs to be flown, the solution was obtained in 1 min. 26 sec. A fleet of DC-10 airplanes with 1,058 planned legs required 6 min. 29 sec. for the solution using a PRIME 550 computer.

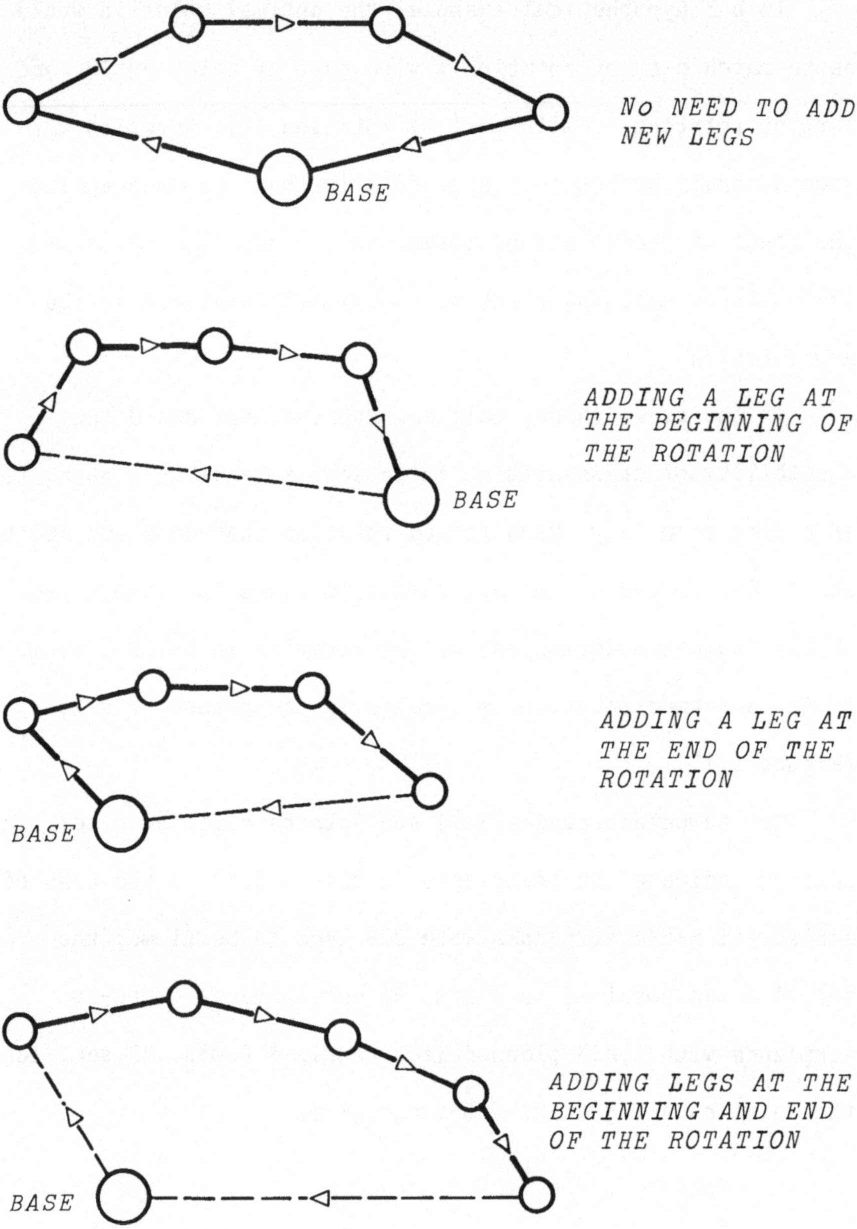

FIGURE 4.17 Tacking on additional legs to formed rotations

4.7. Interactive Computer Programs for Crew Scheduling Problems

Crew scheduling is one of the most important tasks encountered by air carriers. In addition to satisfying different company and/or state regulations having a direct influence on aircraft safety, the crew schedule must also ensure as low air carrier costs as possible. The development of suitable crew scheduling models entails considerable financial outlays and is usually linked to a longer period of time. However, a well-planned crew schedule can provide the air carrier with savings along the line of tens of thousands of dollars a month.

As noted by Gerbracht,[52] it was believed in the 1960s and 1970s that crew scheduling problems of greater dimensions could also be solved by applying operations research techniques and suitable computer technology. Although computer technology has improved considerably, not even the most powerful computer can provide an optimal solution within a reasonable amount of time for larger problems. The previous presentation underscored that different heuristic procedures are used to solve larger problems, and the procedure of Baker et al[9] was described. In recent times, the heuristic procedures being applied are based on interactive computer programs. The formed crew schedule undergoes different changes as it is being carried out. The aircraft rotation plan is given supplemental charter flights, some flights are cancelled while others are late. Clearly, changes in the airline schedule or planned aircraft rotations result in changes in the crew schedule. In such situations, the intuition and experience of the crew planner must be expressed. The computer as an auxiliary aid takes on the main job of different calculations in considering the combinatorial problem and should facilitate the crew planner in making appropriate decisions regarding crew work schedules. This comprises the previous elaboration of interactive computer programs that enable communication between the planner and the computer. Interactive computer programs can be of great assistance to the crew planner in situations when there is a transfer from one airline schedule to another. Interactive programs can also be used in the phase of developing the basic crew rotation plan. Giannessi and Nicolleti[55] indicated the possibility of using interactive computer programs to elaborate a basic crew schedule. In this case, the initial crew schedule would be taken at the start that would not have to satisfy all prescribed constraints. By changing the order of individual flights, the crew planner would continually generate new rotation plans with lower costs than the previous plan, obtaining from the computer information on all parameters that interest him. Gerbracht[52] developed interactive computer programs that improved the crew schedule from step to step. The entire problem is first decomposed into several smaller subproblems. In the next step, interactive computer programs are applied to the smaller subproblems. Ball and Roberts[11] also indicated the possibility of using interactive computer programs in the phase of developing the basic crew schedule, especially in the phase of tacking on additional legs to formed rotations so that all crew rotations start and end in the same base.

Using interactive computer programs requires well trained and experienced crew planners. This considerably facilitates the crew planners' work and enables them to perceive the consequences of all decisions concerning the crew schedule.

4.8. Rostering Crew Members to Planned Work Duties

The previous discussions presented crew scheduling models. Upon completion of the crew schedule, crew members (the pilots and cabin crew) must be assigned to carry out the work duties (rostered). There are frequent cases of a group of 2,000 people who must be assigned to over 20,000 work duties to be carried out during a one-month period. The problem of assigning crew members to planned work duties is combinatorial by nature. Numerous models were developed during the past two decade to assign crew members to work duties, during which time modern computer systems have taken over the manual assignment system. The most important works considering the assignment of crew members to work duties include those of Agard,[2] Nicolleti,[109] Antosik,[4] Buhr,[18] Marchettini,[96] and Glannert.[57] Crew members are usually assigned to work duties to be carried out the following month. This means that the work duties they will be assigned to must be determined for each crew member for every day during the following month. When assigning crew members, consideration must be taken of planned vacations, sick leave. days planned for schooling and training, requested free days, days when medical examinations are held, etc. (Figure 4.18).

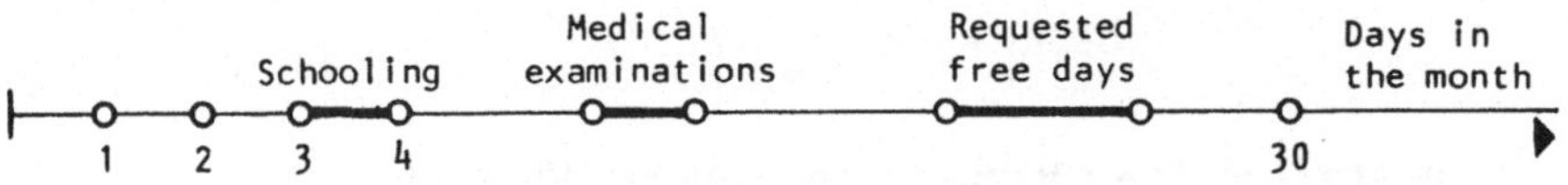

FIGURE 4.18 Days during the month when certain crew members cannot fly

Once the days have been established for each pilot or cabin crew when they cannot be assigned to a work duty, then an approach is made to monthly crew assignments. An important point when developing the monthly assignment is checking the possibility of certain crew members carrying out certain work duties. For example, some crew members as a result of medical or some other reason are not assigned to certain routes. Input data for assignments can be data from the past (one or two years) that indicate how many working days a certain pilot had during that time, how many hours he flew, how many times he worked over the weekend, how many times he landed at unpopular airports, how many times he landed at airports preferred by most pilots, etc. The assignment of pilots and cabin crew to work duties should be done so that all members of the work collective have approximately the same number of free days and working days, approximately the same number of preferred and undesirable rotations, etc. Different air carriers and different authors used various objective functions when finding the optimal monthly crew assignment. For example, Buhr[18] indicates that when finding the optimal monthly crew assignment, the focus can be on minimizing the discrepancy between the average monthly flying time per crew member and the monthly flying time of individual crew members. Buhr[18] also indicates the possibility of minimizing the discrepancy between the average number of days on duty during the month and the number of days on duty of individual crew members. It is also possible to minimize the combination of discrepancy in flying time and the discrepancy regarding days on duty. Antosik [4] indicates minimizing overtime work expenses and minimizing the number of pilots that fly less than the guaranteed amount. The forms of the objective functions will be discussed later.

The problem of developing a monthly crew assignment can be solved using the Hungarian method, the out-of-kilter algorithm or different heuristic procedures. The mathematical problem of developing a monthly assignment can be formulated as follows :

$$\sum_{i=1}^{m} \sum_{j=1}^{n} c_{ij} x_{ij} \rightarrow min \qquad (4.47)$$

$$\sum_{i=1}^{m} x_{ij} = 1 , \quad \forall_j \qquad (4.48)$$

$$\sum_{j=1}^{n} x_{ij} \leqslant 1 , \quad \forall_i \qquad (4.49)$$

whereby :

$$x_{ij} = \begin{cases} 1, & \text{if the } i\text{-th pilot is assigned the } j\text{-th rotation} \\ 0, & \text{otherwise} \end{cases} \qquad (4.50)$$

m - the number of crew members to be assigned,

n - the number of rotations being alloted to the crew members being assigned

$$m \geqslant n$$

c_{ij} - the "costs" of assigning the i-th pilot to the j-th rotation.

Costs c_{ij} of assigning the i-th pilot (or the i-th cabin crew member) to the j -th rotation can comprise different categories. We conclude based on condition (4.48) that each rotation is allotted only one pilot. Expression (4.49) means that the pilot can be assigned to at most one rotation. Inequality (4.51) tells us that the number of crew members being assigned is greater than the number of rotations. Let us discuss this is more detail. In most cases it is not possible to find the optimal solution of assigning pilots or cabin crew to rotations for an entire month in advance. As noted by Antosik, when assigning 60 pilots to 670 rotations, there are 10^{23} possible solutions. Therefore, heuristic procedures are used to compile monthly assignments. Large problems are decomposed into numerous smaller problems and an attempt is made to find the optimal solution for each of the smaller problems. This decomposition can be made in two ways. The first way is called "day by day", and the second "pilot by pilot". The "day by day" method comprises finding a solution for the first day of the month, then the second, etc. The "pilot by pilot" method finds the monthly assignment of the first pilot, then the second, etc. Clear, for the "day by day" method, the number of crew members to be assigned is most often greater than or equal to the number of rotations.

4.9. The "Day-by-Day" Crew Assignment Method

The day-by-day method, as already mentioned, first finds the assignment for the first day of the month, then the second, etc. The algorithm is finished when all crew members are assigned to planned rotations. This method was used in the works of Buhr,[18] Antosik,[4] Marchettini[96] and Glanert.[57]

The day-by-day algorithm is composed of the following steps :

STEP 1 : Note and set aside all rotations that start on the first day of the month.

STEP 2 : Assign crew members to the above rotations. In doing so, make sure that no constraints are breached concerning both the day in question and all previous days for which assignments have already been made.

STEP 3 : Note and set aside all rotations that start on the following day.

STEP 4 : Go to Step 2.

Thus, only those rotations that start on a specific day in the month are set aside from the set of all rotations to be made during the month. Since the dimensions of the problem are considerably smaller, the Hungarian method can be used to solve the problem of crew assignment for a specific day in the month. The Hungarian method was explained in detail in the chapter dealing with long-haul crew scheduling. The main drawback of the day-by-day method lies in the fact that it is very difficult to note problems that might arise when making assignments the following days. It might happen that a larger number of rotations appears for which there are no flight personnel.

4.10. The "Pilot-by-Pilot" Crew Assignment Method

For the pilot-by-pilot method, the monthly assignment is first found for the first pilot, then the second, etc. This method is comprised of the following algorithmic steps :

STEP 1 : Single out a crew member from the list of crew members.

STEP 2 : Assign the crew member to a specific rotation so that all prescribed constraints are satisfied.

STEP 3 : Remove the crew member and all rotations assigned to him from further consideration.

STEP 4 : Go to Step 1.

This procedure is finished when all crew members are assigned to rotations. The first pilot (or cabin crew member) to be considered can be assigned to a certain number of rotations from the set of planned rotations that contains a very large number of elements. For the second pilot, the number of elements in the set of rotations is already somewhat smaller since those rotations assigned to the first pilot have been removed from consideration. The set of rotations corresponding to the pilots that are assigned near the end of the procedure contains a very small number of elements. One result of this is that some rotations are left unassigned. This procedure can also lead to uneven work loads for the flight personnel (unequal monthly flight time, an unequal number of days on duty, etc.).

However, in spite of all the drawbacks, both the day-by-day method and the pilot-by-pilot method enable a monthly assignment of flight personnel that is a good starting basis for crew planners who will make needed corrections to produce the final version of the monthly crew assignment.

4.11 A Combined Heuristic Crew Assignment Method

Noting the drawbacks in the day-by-day and pilot-by-pilot methods, Antosik[4] developed a combined heuristic method to assign crew members to planned rotations. The above two methods are component parts of this new method. Before considering the combination method, we will briefly present a simple heuristic assignment algorithm proposed by Buhr.[18] This heuristic algorithm is composed of the following steps :

STEP 1 : Note and set aside all rotations that start on the first day of the month.

STEP 2 : Of the crew members who have not yet been assigned to a rotation starting on the day in question, note crew member i corresponding to least costs c_{ij} to assign him to any unassigned rotation. Assign rotation j to crew member i . Crew member i and rotation j are excluded from further consideration. By matching crew members and rotations corresponding to least costs, matching continues until all rotations that start on the day in question have been assigned crew members.

STEP 3 : Go to the next day and repeat the procedure described in Step 2.

In addition to this simplified heuristic algorithm, Buhr[18] proposed an approximation method to assign crews to rotations. However, computer testing indicated that the proposed approximation method does not give better results than those from the above heuristic algorithm both in terms of the quality of the solution and in terms of needed computer time.

Let us return to the combined heuristic method proposed by Antosik.[4] This method, developed within AIR FRANCE, is called the CADET method. The main focus of CADET is to "cover" all planned rotations with specific crew members. The second goal which was of lower priority, was to reduce disproportions in the flight duties between individual crew members to the lowest possible degree. (Disproportions in terms of monthly flight time, number of days on duty, number of preferred rotation assignments, number of undesired rotation assignments, etc.). As mentioned earlier, CADET is a combination of the pilot-by-pilot method and the day-by-day method. First, the pilot-by-pilot method is applied in which every pilot (or cabin crew personnel) is assigned as many rotations as possible. To this effect, the greatest care is given to satisfying the first goal of covering as many rotations as possible with corresponding crew members. Upon completion of this step there are often considerable disproportions in terms of monthly flight time, days on duty and other parameters. In the next step, CADET applies the day-by-day method in order to reduce these disproportions. It is important to note that all rotations that were assigned after the first step must remain assigned after applying the day-by-day method. The goal of the second step is only to even out the flight personnel's work load.

Before applying the pilot-by-pilot method, all pilots should be designated by a corresponding ordinal number depending on the number of rotations to which the pilot can be assigned. Ordinal number 1 denotes a pilot who can be engaged on the least number of rotations, while the last ordinal number denotes a pilot who can be engaged on the largest number of rotations. This system of issuing ordinal numbers maintains the emphasis on covering the largest number of rotations with pilots. When applying the pilot-by-pilot method, every pilot is assigned a rotation that is as long as possible and the emphasis is on every pilot being assigned to the largest number of rotations possible.

When applying the pilot-by-pilot method, each pilot is assigned rotations to be flown during the next 30-day period (Figure 4.19), so that data on the number of flight hours, day on duty, etc., are known for each pilot. The day-by-day method is applied for every day in the month. The first day is started with and an attempt is made to exchange rotations between pilots in order to even out their work load.

Let us take the third day as an example. Since each day is taken separately, we note that for the first and second day pilot P_1 has been assigned to rotations R_1 and R_2. And we will consider rotation R_4 to be already assigned to pilot P_1 although the fourth day has not yet undergone the day by day method. So for the third day we must investigate

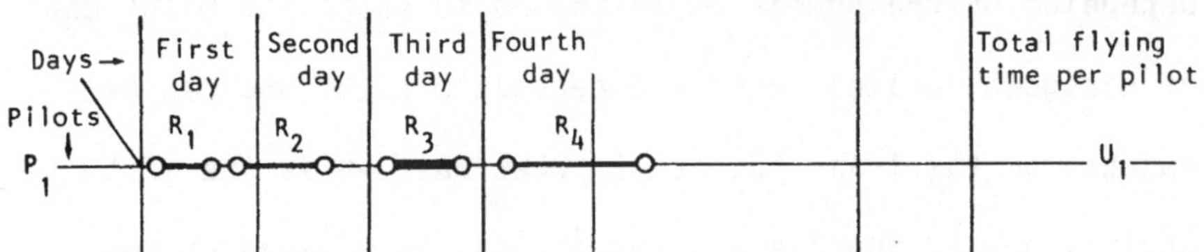

FIGURE 4.19 Rotations assigned to pilots by the pilot-by-pilot method

whether rotation R_3 should be given to pilot P_1 or some other pilot.

The set of rotations assigned to pilot P_1 is denoted by T_1. This set contains elements R_1, R_2 and R_4, i.e. $T_1 = \{R_1, R_2, R_4\}$ Pilots P_2, P_3, P_4 and P_5 respectively are assigned sets of rotations $T_2 = \{R_5, R_6\}$, $T_3 = \{R_9, R_{10}, R_{11}\}$

$T_4 = \{R_{12}, R_{13}, R_{16}\}$ and $T_5 = \{R_{17}, R_{18}, R_{19}, R_{20}, R_{21}\}$. Pilots must be chosen for rotations R_3, R_7, R_8, R_{14} and R_{15} (denoted on Figure 4.19 by the heavy line) in order to even up each pilot's flying time as much as possible. This problem is solved by the Hungarian or some other method. We denote by F_{ij} the costs of matching rotation R_j with the set of rotations T_i that has been assigned to a specific pilot. Antosik[4] proposed to calculate costs F_{ij} as :

$$F_{ij} = \begin{cases} |S_{ij} - S^*|, & \text{if adding rotation } R_j \text{ to the set of already assigned rotations } T_i \text{ does not go against constraints} \\ \infty, & \text{if adding rotation } R_j \text{ to the set of assigned rotations } T_i \text{ goes against constraints} \end{cases} \quad (4.52)$$

where :

S_{ij} - the total flight time of pilot P_i when rotation R_j is added to the set of already assigned rotations T_i,

S^* - average pilot flight time.

Taking expression (4.52) into consideration, we conclude that the set of already assigned rotations should be expanded to include new rotations so as to minimize the total discrepancy between the flight times of some pilots and the average flight time per pilot. Variables F_{ij} relating to the example in question are given in Table 4.5.

TABLE 4.5

	R_3	R_7	R_8	R_{14}	R_{15}
T_1	$F_{1,3}$	∞	∞	**$F_{1,14}$**	$F_{1,15}$
T_2	$F_{2,3}$	$F_{2,7}$	**$F_{2,8}$**	$F_{2,14}$	$F_{2,15}$
T_3	**$F_{3,3}$**	∞	∞	$F_{3,14}$	$F_{3,15}$
T_4	$F_{4,3}$	∞	∞	$F_{4,14}$	**$F_{4,15}$**
T_5	$F_{5,3}$	**$F_{5,7}$**	∞	$F_{5,14}$	$F_{5,15}$

The optimal matching is obtained by applying the Hungarian (or some other) method. The circled values of F_{ij} result in the optimal matching. Upon finding the optimal matching of rotations that were unassigned with the sets of rotations that are already assigned, there is a redistribution of rotations between some pilots. For example, pilot P_1 after applying the pilot-by-pilot method will exchange rotation R_3 that he should have flown for rotation R_{14} that should have been flown by pilot P_4. Rotation R_3 that should have been flown by pilot P_1 will be flown by pilot P_3, etc. (see Table 4.5 and Figure 4.19).

The final monthly flight personnel assignment is obtained when the pilot-by-pilot method is followed by the day-by-day method, with the Hungarian method being applied for each day in the month starting with the first day. The CADET method and other models for monthly crew assignments have numerous advantages over manual monthly assignments. Many authors have shown that utilization of a suitable model and a modern computer system provide much better results with minimal disproportions in terms of the flight personnel's work load. With a computer, it is possible to assign the monthly schedule the last week of the present month for the following month when relatively trustworthy input data is available. For the manual method, the procedure usually starts already the first week of the previous month when precise enough input data are not available. As noted by Antosik,[4] computer time in the CADET method is only 3-4 seconds per crew member.

4.12 Calculating the Costs of Assigning Cabin Crew to Planned Rotations

The problem of elaborating the monthly crew assignment is defined by expressions (4.47) - (4.51). Figuring in the objective function that is minimized are costs c_{ij} comprising the cost of assigning the i -th pilot (or cabin personnel) to the j -th rotation.

Buhr[18] called costs c_{ij} "malus" and calculated them by taking into consideration the cumulative number of flying hours and the cumulative number of days that individual crew members spend on duty. Buhr [18] calculated these costs as follows.

The day in the month in which the rotations in question end is denoted by M_j. The number of days in the month in question is denoted by M. The number of pilots (or cabin crew member) that we want to assign to the rotation is denoted by Z, and F denotes the total number of hours that will be flown by the group of pilots in question during the entire month. The average number of monthly flight hours per pilot y equals :

$$y = \frac{F}{Z} \tag{4.53}$$

The cumulative number of hours that an individual pilot flies from the beginning of the month to a certain day in the month is denoted by FH_i. The optimal value of this number of flight hours OFH equals :

$$OFH = y \cdot \frac{M_j}{M} = \frac{F}{Z} \cdot \frac{M_j}{M} \tag{4.54}$$

Since the average number of monthly flight hours per pilot equals $\frac{F}{Z}$, we conclude that ending with day M_j, the pilot in question flew $\frac{F}{Z} \cdot \frac{M_j}{M}$ which equals the average flight time (Figure 4.20).

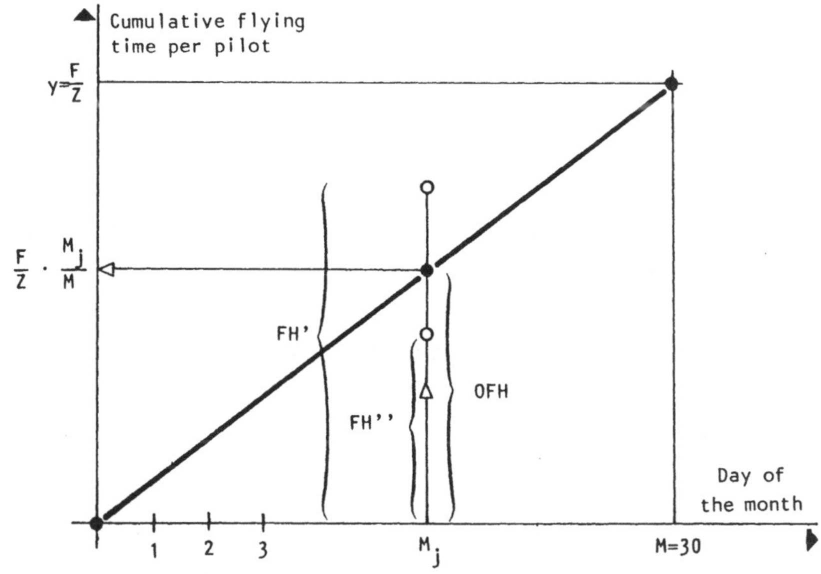

FIGURE 4.20   Cumulative flight time per pilot

If the pilot in question flew FH' hours, we conclude that he would then fly more than the optimal amount (Figure 4.20). Flight time FH'' is less than optimal. Since we are endeavoring to minimize the disproportion in terms of crew work load, it is clear that assigning a pilot to a rotation is all the better if the flight time is near the optimal (average) time. Buhr [18] mentioned that Lufthansa guaranteed its pilots 11 days of duty per month. Therefore, the optimal cumulative number of days that the pilot spends on duty ending with day OFD equals :

$$OFD = \frac{M_j}{M} \cdot 11 \qquad (4.55)$$

Buhr [18] proposed that costs c_{ij} of assigning pilot i to rotation j be calculated using this expression :

$$c_{ij} = (FH_i + RH_j - OFH) \cdot W_1 + (OFD - FD_i - RD_j) \cdot W_2 \qquad (4.56)$$

where :

FH_i - total number of hours flown by pilot i from the beginning of the month,

RH_j - number of flight hours contained in rotation j ,

OFH - average number of monthly flight hours per pilot,

OFD - average number of monthly days on duty per pilot,

FD_i - total number of days on duty by pilot i since the beginning of the month,

RD_j - number of duty days contained in rotation j,

W_1, W_2 - weight factor associated with the flight,i.e. number of days on duty.

The weight factor values are given by the crew planner. When $W_1 > W_2$ priority is given to evening the number of hours flown, while with $W_2 > W_1$ priority is given to evening the number of days on duty. Values c_{ij} are calculated for all pairs of pilots and rotations for every day in the month represent the input data for applying the day-by-day method, or for applying some other method.

Nicoletti also used the weight factor and the linear function between cumulated flight time (the cumulative number of days on duty and other parameters) and days in the month. Marchettini[96] also used a similar procedure to calculate costs c_{ij}.

4.13 Determining the Number of Reserve Crew

In addition to forming the crew schedule and the monthly assignment of crew members, the number of reserve crew needed must also be calculated. When executing the planned airline schedule, perturbations arise from time to time in traffic, crew members are suddenly taken ill, there are unplanned delays on the way to work duty, etc., giving rise to the need for reserve crew members to take the place of the planned crew. A number of crew must be in reserve every day in order to step in where needed. The number of such reserve crew members must be adequately established. In addition to determining the needed number of reserve crew, the problem also arises of how to use them during different periods of the day. The problem of determining needed reserve crew was discussed by Bornemann,[16] Haydon,[64] Gaballa,[41] and Mitchell.[105] Gaballa also considered tactics on using reserve crew during the day. Too many reserve crew considerably increase crew costs. On the other hand, a large number of reserve crew improve the exactitude of traffic due to the small number of fights that can be cancelled or shifted to the next day due to the air carrier's lack of crew for a planned flight. Too small a number of reserve crew can cause a rise in the number of cancelled flights and a rise in the air carrier's costs associated with such cancellations (hotel accommodations and other compensations to the passengers of cancelled flights). Considering the size of crew expenses and the fact that costs per cancelled international flight can be over $50,000, it is clear that considerable attention must be given to determining the number of reserve crew needed and their assignment to duty during the day.

The basic input data to calculate the number of reserve crew needed are statistical data concerning the number of requests for reserve crew over a certain period of time. The intensity with which reserve crew are called to work can be calculated based on these data. Haydon[64] used previous statistical analyses and added to each takeoff a probability equal to 0.025 that there would be a need for a reserve crew in order to finish the planned flight. Let us denote this probability by p. We denote by n the number of takeoffs that should be made during a certain period of time. Probability p that a reserve crew will be needed is associated with every takeoff. The probability that a reserve crew will be called to make a specific flight does not depend on whether the call arises during the previous takeoffs or whether it will arise during the next takeoff. We denote by X the number of possible calls for reserve crews ($X=0,1,2,\ldots,n$). Probability $P^{(n)}(X=x)$ that during n takeoffs there will be calls for reserve crew equals :

$$P^{(n)}(X=x) = \binom{n}{x} (p)^{x} (1-p)^{n-x} \tag{4.57}$$

The probability distribution determined by (4.57) is called a binomial probability distribution. Gaballa,[41] in addition to Haydon,[64] used the binomial probability distribution when determining the number of needed reserve crew.

The number of hours that a reserve crew can spend on duty waiting for a possible call is also regulated by company rules. If no reserve crew exists, the probability of a flight being postponed to the next day equals :

$$P\begin{pmatrix}\text{postponing}\\ \text{flight to}\\ \text{next day}\end{pmatrix} = P\begin{pmatrix}\text{at least}\\ \text{one request}\\ \text{for reserve}\\ \text{crew}\end{pmatrix} = P^{(n)}(X>0) =$$

$$= 1-P^{(n)}(X=0)=1-\binom{n}{o}p^{o}(1-p)^{n}=1-n!(1-p)^{n} \tag{4.58}$$

If there is only one reserve crew, the probability of a flight being postponed to the following day equals :

$$P\begin{pmatrix}\text{postponing}\\ \text{flight to}\\ \text{next day}\end{pmatrix} = P\begin{pmatrix}\text{more than}\\ \text{one request}\\ \text{for reserve}\\ \text{crew}\end{pmatrix} = P^{(n)}(X>1) =$$

$$= 1-P^{(n)}(X=0)-P^{(n)}(X=1)=1-n!(1-p)^{n}-np(1-p)^{n-1} \quad (4.59)$$

The probabilities calculated based on expressions (4.58) and (4.59) refer to the period of time when a reserve crew is on duty. As the period of time during which a reserve crew is on duty is considerably shorter than the period of time during which flights are made during the day, this calculation of the probability that a flight will be postponed must refer to the entire day. Haydon [64] and Gaballa [41] provided examples in their papers that concern the entire day.

Figure 4.21 denotes planned takeoffs during a day as well as the beginning and ending working time of reserve crews.

In the example we are considering, there are 18 planned takeoffs (takeoff times are marked with black dots). The reserve crew work 12 hours.

We denote by p_i^c the probability of the i-th planned takeoff being cancelled (by cancellation we mean its postponement to the next day). The first flight is planned to takeoff at 5:30 and it cannot be postponed to the next day since any cancellation of the planned crew will result in the reserve crew taking over that is on duty at 5:00. This means that :

$$p_1^c = 0 \quad (4.60)$$

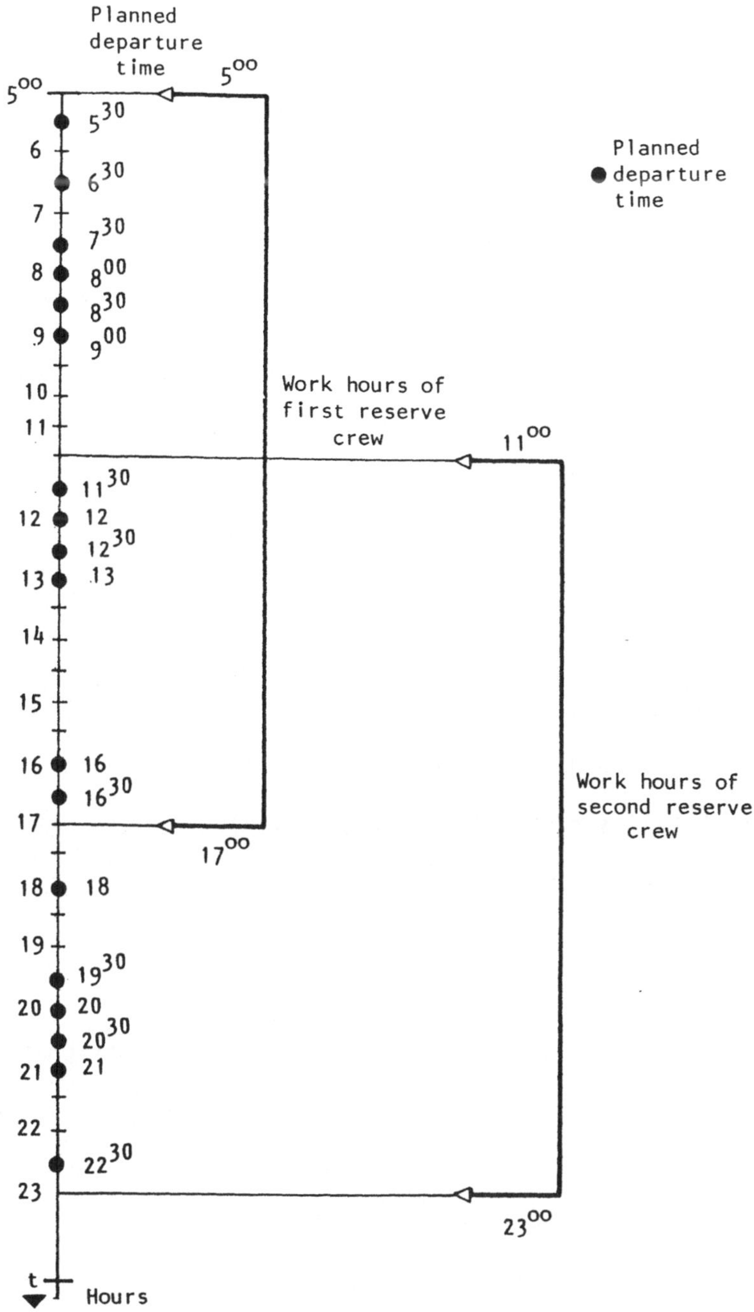

FIGURE 4.21 Assigning reserve crew to duty during the day

The second flight cannot be postponed to the next day, either. If there is no request for the reserve crew during the first flight, and there is during the second, then the reserve crew can take on the work duty of the second flight. If both the first and the second takeoffs requested reserve crews, the second flight can be made by the crew coming on duty at 11:00. The second takeoff would be late (11:00 instead of 6:30), but it would not be postponed to the following day. We conclude that the probability of postponing the second flight to the following day equals zero, i.e. :

$$p_2^c = 0 \tag{4.61}$$

The third flight can be postponed until the next day if the first, second and third takeoffs request reserve crews. The probability of the third flight being postponed until the next day equals :

$$p_3^c = \binom{3}{3} p^3 (1-p)^0 = p^3 \tag{4.62}$$

The probability that the fourth flight will be postponed until the next day equals the product of the probability that the fourth flight will request a reserve crew and the probability that the number of requests for reserve crews on the first, second and third flights is greater than 1. This means that :

$$p_4^c = p\cdot P^{(3)}(X>1) = p\cdot[1-P(X=0)-P(X=1)] =$$
$$= p\cdot[1-\binom{3}{0}p^0(1-p)^3-\binom{3}{1}p^1(1-p)^2] =$$
$$= p\cdot[1-(1-p)^3-3p(1-p)^2] \quad (4.63)$$

We similarly calculate the probability of the fifth, sixth,..., twelfth and thirteenth flights being postponed until the next day.

The probability that the fourteenth flight that should takeoff at 19:30 is postponed to the next day equals :

$$p_{14}^c = p\cdot P^{12}(X>1)+p\cdot[1-P^{12}(X>1)]\cdot P^1(X>0) \quad (4.64)$$

The first addend in the above expression represents the probability that there will be a request for a reserve crew on the fourteenth flight with more than one reserve crew request on the first twelve takeoffs. The second addend represents the probability that there is a request for a reserve crew on the fourteenth takeoff when not more than one reserve crew was requested on the first twelve takeoffs, with the thirteenth takeoff requesting a reserve crew. The probability of postponement until the next day for the fifteenth, sixteenth, seventeenth and eighteenth flights is calculated in a similar fashion.

We denote the cost of postponing the i-th flight until the next day by C_i. The mean cost of postponing flights to the next day equals :

$$C = \sum_{i=1}^{n} C_i \cdot p_i^c \quad (4.65)$$

The mean number of flights during the day that will be postponed until the next day equals :

$$N = \sum_{i=1}^{n} 1 \cdot p_i^c = \sum_{i=1}^{n} p_i^c \quad (4.66)$$

The mean time between the postponement of two flights to the next day equals :

$$\bar{t} = \frac{1}{N} = \frac{1}{\sum_{i=1}^{n} p_i^c} \quad (4.67)$$

Knowing the costs of engaging a reserve crew defined by expressions (4.65), (4.66) and (4.67) enables suitable decisions to be reached concerning the number of needed reserve crew. Haydon[64] demonstrated in his example that with a rise in the number of reserve crew from 2 to 3, the mean time between two postponed flights increases from 69 days to 22 months. Introduction of a fourth reserve crew increases this time to 23 years. Gaballa[41] indicated that changing the number of reserve crew and varying the tactics of their utilization (different starting and finishing work hours) can realize considerable financial savings.

5. SEAT RESERVATION SYSTEMS

5.1. Characteristics of the Seat Reservation Process

Seats are traditionally reserved in advance for a specific flight. In most cases, passengers can cancel their reservations without any consequences. Also, a certain number of passengers do not appear for the flight on which they have reserved a seat, neither do they cancel this reservation. These "no-show" passengers usually do not bear any financial consequence for their behavior. If the air carrier accepts only as many reservations as there are seats in the aircraft, upon departure a certain number of seats will remain empty due to cancellations, resulting in a lower load factor and lower profit. For this reason, air carriers usually accept more reservations than there are seats in the aircraft in order to decrease the number of empty seats on takeoff. However, when the air carrier accepts more reservations than the aircraft's carrying capacity, more passengers might appear than there are seats in the plane. It is possible that a passenger with a normally accepted reservation cannot find a seat on his desired flight. The level must therefore be established to which reservations will be accepted during a certain period of time.

The seat reservation system has an essential effect on the air carrier's economic results (expressed through the average load factor) and also has a direct influence on the quality of the transportation service (measured by the number of passengers turned away with normal reservations) and on the air carrier's reputation. It is important to underscore that efforts put into developing and implementing seat reservation models absolutely have their economic justification., As noted by Rothstein,[135] a 1% increase in the average passenger load results in an annual rise in the air carrier's profits measured in the millions of dollars. The importance of seat reservation problems has resulted in the development of numerous models over the past two decades. Some of the more important authors dealing with this topic are Rothstein and Stone,[131] Simon,[141] Falkson,[39] Simon,[142] Martinez and Sanchez,[100] Rothstein,[132,133] Vickrey,[167] Littlewood,[93] Andersson,[3] Shlifer and Vardi,[145] Rothstein,[13] Mayer,[102] Gasco,[45] Nagarajan,[107] Gerbracht,[50] Gasco,[46] Wang,[170] Titze and Griesshaber[164] and Rothstein[135] Most of these authors are from airline companies, thus some of these models have been verified in practice. The AGIFORS organization (Airline Group of the International Federation of Operational Research Societies) that deals with applied methods of operations research in airline companies formed a special study group on the problem of seat reservations as early as 1964.

The air carrier's reservation system is composed of a computer, a network of agents and a telecommunications network. One of the basic questions regarding the reservation system is its centralization. There is no simple answer to the question as to whether or not the reservation system should be centralized. A decision in this regard should only be made after a detailed technical-economic study that estimates the costs of personnel, telecommunications, service quality, etc., for alternative reservation systems. Special attention should be given to providing the telecommunications capacities with accurate dimensions and reliable equipment.

The reservation system is rendered more complicated by the air fare system. There are usually several different air fares for the same route. In some cases the difference between the least expensive and the most expensive fare can be quite significant. Often passengers paying the lower fare will reserve seats before passengers paying the higher fare, resulting in passengers paying the higher fare not being able to find a seat on their desired flight. This in itself reduces the air carrier's profit. Determining the reservation level for passengers with lower fares is also a problem that must be given considerable attention as its accurate solution can result in significant economic effects. The problem is even more complex when dealing with the reservation level of passengers with lower fares when the flight has several legs.

The development and implementation of models dealing with different aspects of the reservation system must be given due attention in order to achieve satisfactory economic effects and offer the passengers suitable transportation services. The following discussions will treat specific reservation models in more detail.

5.2. The Optimal Reservation Level for Nonstop Flights with One Type of Air Fare

As mentioned earlier, passengers can most often cancel their reservations or not even show up for the flight, with little or no consequences. In order to improve the aircraft's load, air carriers usually accept more reservations than the aircraft's capacity, counting on a certain number of passengers to cancel their reservations. Due to the probabilistic nature of the process of reserving a seat and cancelling a reservation, it might happen that a passenger with an accepted reservation cannot find a seat on his desired flight. For many years, acceptance of more reservations than the aircraft's capacity was considered to be one of the air carrier's business secrets. For reasons of prestige, air carrier's did not want to admit that they accepted more reservations than the aircraft capacity, thereby making it impossible in certain situations for some passengers with reservations to obtain a seat on their desired flight. For a time, airline companies paid certain compensations to the overbooked passengers. Passengers who did not appear for the flight often paid the same amount. This policy led to misunderstandings between passengers and air carriers. To this effect, passengers who could not appear for the flight due to objective reasons (previous flights that were late or cancelled, etc.) could not be distinguished from passengers who did not appear for strictly subjective reasons. In addition, some passengers were not satisfied with the compensations offered to them by air carriers when they could not find a seat on their reserved flight. The case of American lawyer Ralph Nader is well known : he received a court decision of $25,000 in damages from Allegheny Airlines in 1972 when he could not get a seat on his reserved flight (Rothstein [135]). Many authors (Simon, [141] Falkson, [39] Rothstein [131] Vickrey [167]) discussed various possibilities for compensating overbooked passengers.

We will go into the possibility of solving this problem in a satisfactory manner later on. We would only mention that to date no solution has been implemented that completely satisfies both passengers and air carriers. Many companies use waiting lists as a means of solving the complex reservation problem. Reservations are accepted for a certain flight up to a preordained reservation level (that ia equal to or greater than the aircraft's carrying capacity). New passengers wanting to make a reservation are put on a waiting list if this previous reservation level has already been met. In case of a cancellation, the first passenger on the waiting list receives the reservation, the second one on the waiting list becomes the first on the list, etc. However, it is often impossible to inform waiting list passengers of their change in status. In addition, putting a waiting list passenger on the aircraft instead of some passenger who did not appear assumes the physical presence of the waiting list passenger at the airport, whereby waiting list passengers know that there is only a possibility of getting a seat on their desired flight. In any case, introduction of waiting lists cannot completely solve the complex problems of the reservation process.

Let us consider the problem of establishing the reservation level in a nonstop flight with only one type of air fare. Figure 5.1 presents the typical dependence between the total number of reservations and time before takeoff.

The total number of reservations usually rises as departure time approaches. From time to time there are smaller drops in the total number of reservations due to cancellations. Let us first consider the case when only one reservation level exists during the entire time reservations are taken. The higher the reservation level, the greater the

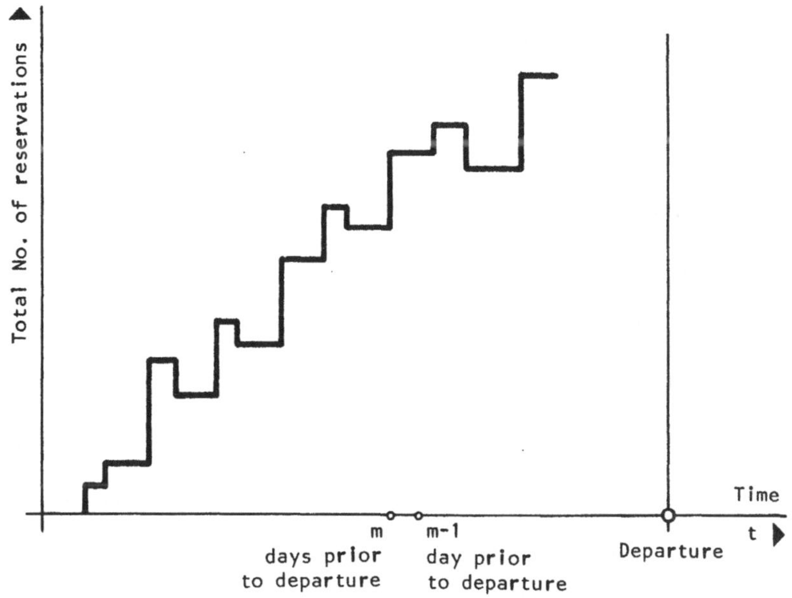

FIGURE 5.1 Dependence between total number of reservations and time before departure

danger of a reserved passenger appearing and not being able to find a seat in the plane. The lower the reservation level the greater the number of seats that can remain empty upon departure. The optimal reservation level can be obtained as the solution to an optimization problem that strives to maximize the air carrier's profit by taking into account both revenue from passengers sitting in the aircraft and compensation that must be paid to reserved passengers without seats. Gerbracht [50] developed a model to determine the reservation level that maximizes air carrier profit when overbooked passengers are paid a compensation. We will briefly present the results of this model. Gerbracht first of all considered the problem with the assumption that the number of no-show passengers does not depend on the number of reservations. Clearly, this number must depend on the number of reservations since, for example, a greater number of no-show passengers can be expected when there are 120 reservations than when there are only 80 reservations.

However, we assume, anyway, that the number of no-show passengers does not depend on the number of reservations. Later we will discuss the more complex case when these passengers are assumed to be a function of the number of reservations. We denote by :

n - the aircraft capacity (number of seats),

C - average air fare on the flight,

P - average compensation paid to overbooked passengers,

L - reservation level.

The probability that k passengers will not show up for departure is denoted by $p(k)$.

If $L \leqslant n$ the number of reservations is less than the number of seats in the aircraft and there is no danger of any overbooked passengers. The number of passengers that will not appear at takeoff k , lies within boundaries :

$$0 \leqslant k \leqslant L \qquad (5.1)$$

When all passengers with reservations appear at takeoff, the air carrier will have revenues equal to $L \cdot C$. Should one passenger cancel, revenues equal $(L-1)C$, for two cancellation, revenues equal $(L-2) \cdot C$, etc. Since the number of passengers that will not show up is a random variable, then revenues are also a random variable. Since these revenues are a function of reservation level L, we denote it by $R(L)$. Mean revenues $\overline{R}(L)$ equal :

$$\overline{R}(L) = \sum_{k=0}^{L} (L-k) \cdot C \cdot p(k) \qquad (5.2)$$

i.e. :

$$\overline{R}(L) = C \left[L \sum_{k=0}^{L} p(k) - \sum_{k=0}^{L} kp(k) \right] \qquad (5.3)$$

Since :

$$\sum_{k=0}^{L} p(k) = 1 \quad i \quad \sum_{k=0}^{L} kp(k) = \overline{k} \qquad (5.4)$$

with $\overline{k}$ the mean number of passengers that will not appear at takeoff, this is :

$$\overline{R}(L) = C(L - \overline{k}) \qquad (5.5)$$

Based on expression (5.5) we can conclude that when $L \leqslant n$ revenues are greater the greater the reservation level.

Let us now discuss the case when $L > n$. If all passengers who made a reservation appear $(L-n)$ some passengers will not get a seat. The air carrier in this case will earn revenue by transporting n passengers and will have to pay a monetary compensation equal to P to each of the $(L - n)$ remaining passengers. In other words, revenues equal :

$$n \cdot C - (L - n) \cdot P$$

When one passenger does not show up at departure, the air carrier must pay compensation to $(L-n-1)$ passengers, so that revenues equal :

$$nC - (L - n - 1) P$$

When two passengers do not appear at takeoff, revenues are :

$$nC - (L - n - 2) \cdot P$$

When $(L-n)$ passengers do not show up, then no passenger must be paid a compensation and revenues equal :

$$n C$$

Revenues equal $(n-1) \cdot C$ when $(L-n+1)$ passengers do not show up, etc. The mean realized revenues $\overline{R}(L)$ when $L > n$ equals :

$$\overline{R}(L) = \sum_{k=0}^{L-n} \left[nC-(L-n-k) \cdot P\right] \cdot p(k) + \sum_{k=L-n+1}^{L} (L-k)C \cdot p(k) \qquad (5.6)$$

By adding and subtracting expression $\sum_{k=0}^{L-n} (L-k)C \cdot p(k)$

we get :

$$\overline{R}(L) = \sum_{k=0}^{L-n} [nC-(L-n-k)P]p(k) + \sum_{k=L-n+1}^{L} (L-k)Cp(k) +$$

$$+ \sum_{k=0}^{L-n} (L-k)Cp(k) - \sum_{k=0}^{L-n} (L-k)Cp(k) \qquad (5.7)$$

i.e. :

$$\overline{R}(L) = \sum_{k=0}^{L} (L-k)Cp(k) + \sum_{k=0}^{L-n} nCp(k) -$$

$$- \sum_{k=0}^{L-n} [(L-n-k)P]p(k) - \sum_{k=0}^{L-n} (L-k)Cp(k) \qquad (5.8)$$

$$\overline{R}(L) = \sum_{k=0}^{L} (L-k)Cp(k) + \sum_{k=0}^{L-n} nCp(k) - \sum_{k=0}^{L-n} [(L-n-k)P]p(k) -$$

$$- \sum_{k=0}^{L-n} (L-n-k)Cp(k) - \sum_{k=0}^{L-n} nCp(k) \qquad (5.9)$$

Finally :

$$\overline{R}(L) = \sum_{k=0}^{L} (L-k)Cp(k) - (C+P)\sum_{k=0}^{L-n} (L-n-k)p(k) \qquad (5.10)$$

i.e. :

$$\overline{R}(L) = LC \sum_{k=0}^{L} p(k) - LC \sum_{k=0}^{L} kp(k) -$$

$$-(C+P) \sum_{k=0}^{L-n} (L-n-k)p(k) \qquad (5.11)$$

$$\overline{R}(L) = (L-\overline{k})C-(C+P) \sum_{k=0}^{L-n} (L-n-k)p(k) \qquad (5.12)$$

Our problem lies in choosing the value of variable L **that maximizes mean revenues** $\overline{R}(L)$ **. In order to determine this value, we first establish how mean revenues change with a rise in value** L **. We determined the difference between values** $\overline{R}(L+1)$ **and** $\overline{R}(L)$ **. This is :**

$$\overline{R}(L+1)-\overline{R}(L) = (L+1-\overline{k})C-(C+P) \sum_{k=0}^{L+1-n} (L+1-n-k)p(k) -$$

$$- (L-\overline{k})C+(C+P) \sum_{k=0}^{L-n} (L-n-k)p(k) \qquad (5.13)$$

$$\overline{R}(L+1)-\overline{R}(L) = C-(C+P) \sum_{k=0}^{L-n} (L+1-n-k)p(k) -$$

$$- (C+P)(L+1-n-L-1+n)p(L+1-n) +$$

$$+ (C+P) \sum_{k=0}^{L-n} (L-n-k)p(k)$$

i.e. :

$$\overline{R}(L+1)-\overline{R}(L) = C-(C+P)\sum_{k=0}^{L-n}(L+1-n-k-L+n+k)p(k) \quad (5.14)$$

$$\overline{R}(L+1) - \overline{R}(L) = C - (C+P) \sum_{k=0}^{L-n} p(k) \quad (5.15)$$

If $\overline{R}(L+1)-\overline{R}(L) > 0$ it is worth our while to increase the reservation level by 1. Increasing the reservation level brings us greater and greater revenues until the difference between $\overline{R}(L+1)-\overline{R}(L)$ is positive. Thus the optimal value for L is the first value for which :

$$\overline{R}(L+1) - \overline{R}(L) \leqslant 0 \quad (5.16)$$

Based on expressions (5.15) and (5.16) we conclude that the optimal value for L is obtained by solving this inequality :

$$C - (C+P) \sum_{k=0}^{L-n} p(k) \leqslant 0 \quad (5.17)$$

We denote by L^* the optimal value of L. L^* is the smallest value for which :

$$C - (C+P) \sum_{k=0}^{L^*-n} p(k) \leqslant 0 \quad (5.18)$$

i.e. :

$$\sum_{k=0}^{L^*-n} p(k) \geqslant \frac{C}{C+P} = \frac{1}{1+\frac{P}{C}} \qquad (5.19)$$

The number of passengers that will not appear at departure k is a random variable that takes an integer value from the interval $[0,L]$. We denote by $F(z)$ the cumulative density function of this random variable that is defined as :

$$F(z) = \sum_{k=0}^{z} p(k) \qquad (5.20)$$

We can conclude that the optimal reservation level L^* is the smallest value greater than n for which :

$$F(L^*-n) \geqslant \frac{1}{1+\frac{P}{C}} \qquad (5.21)$$

The optimal reservation level is determined based on expression (5.21) that was obtained by Gerbracht [50] for the well-known cumulative density function of the number of no-show passengers. The number of passengers with reservations that cannot get a seat on their desired flight is also a random variable. When all passengers appear the number of passengers turned away is $L-n$. When one passenger does not appear the number turned away is $L-n-1$. When two passengers do not appear the number turned away is $L-n-2$, etc. The mean number of passengers turned away $\bar{Q}(L)$ equals :

$$\bar{Q}(L) = (L-n)\cdot p(0) + (L-n-1)\cdot p(1) + (L-n-2)\cdot \cdot p(2)+ \ldots + 1\cdot p(L-n-1) + 0\cdot p(L-n) \qquad (5.22)$$

i.e. :

$$\overline{Q}(L) = \sum_{k=0}^{L-n} (L-n-k)\cdot p(k) \qquad (5.23)$$

The distribution of no-show passengers, i.e. the mean of this distribution, has a great influence both on the optimal reservation level and on the mean number of passengers turned away with reservations. Thus suitable statistical data must be very selectively gathered on the number of no-show passengers. In some, rare cases the number of no-shows can be very large either because of certain delayed or late flights on which passengers were located or due to some other reason. Statistical data referring to such situations should not be part of the analysis so as not to result in an estimated mean number of no-shows that is greater than it really is.

In the previous discussion, we determined the optimal reservation level and mean number of passengers turned away with reservations assuming that the number of no-show passengers did not depend on the number of reservations. It is more realistic to assume that these two values are dependent on each other. A greater number of no-show passengers can certainly be expected when there is a greater number of reservations. Let us first discuss the situation, as we did earlier, when $L \leqslant n$, i.e. when the reservation level is less than the number of seats in the aircraft. Mean revenues $\overline{R}(L)$ equal :

$$\overline{R}(L) = \sum_{k=0}^{L} (L-k) \cdot Cp(k) \qquad (5.24)$$

We denote by X the random value representing the share of no-show passengers out of total passengers, i.e. :

$$k = X \cdot L \qquad (5.25)$$

whereby $0 \leqslant X \leqslant 1$.

The probability density function of random variable X is denoted by $f(x)$.Based on expressions (5.24) and (5.25) we get :

$$\overline{R}(L) = \int_0^1 (L-xL)Cf(x)dx = LC \int_0^1 (1-x)f(x)dx \qquad (5.26)$$

$$\overline{R}(L) = LC \int_0^1 f(x)dx - LC \int_0^1 xf(x)dx \qquad (5.27)$$

i.e. :

$$\overline{R}(L) = LC - LC\overline{x} = LC(1-\overline{x}) = C(L-\overline{k}) \qquad (5.28)$$

whereby :

$\overline{x} = \int_0^1 xf(x)dx$ - mean share of the number of no-show passengers out of total passengers

$\int_0^1 f(x)dx = 1$

$\overline{k} = \overline{x}L$ - mean number of no-show passengers

Based on expression (5.28) we conclude that we will realize greater revenues the greater the reservation level when $L \leqslant n$.

When the reservation level is greater than the number of seats in the aircraft, i.e. when $L>n$, there will be a certain number of passengers turned away with reservations. As in the previous discussion, we assume that the air carrier must pay damages to each of these passengers with P financial units. We earlier showed (expression 5.16) that revenues in this case are :

$$\overline{R}(L) = (L-\overline{k})C - (C+P) \sum_{k=0}^{L-n} (L-n-k)p(k) \qquad (5.29)$$

Since $k=xL$, this is :

$$\overline{R}(L) = C(L-\overline{k})-(C+P) \int_{0}^{1-\frac{n}{L}} (L-n-xL)f(x)dx \qquad (5.30)$$

i.e. :

$$\overline{R}(L) = C(L-\overline{k})-(C+P) \int_{0}^{1-\frac{n}{L}} \left[L(1-x)-n\right] f(x)dx \qquad (5.31)$$

$$\overline{R}(L) = CL(1-\overline{x})-(C+P) \int_{0}^{1-\frac{n}{L}} \left[L(1-x)-n\right]f(x)dx \qquad (5.32)$$

By finding derivative $\frac{d[\overline{R}(L)]}{dL}$ and setting it equal to zero, we get :

$$\frac{d[\overline{R}(L)]}{dL} = C(1-\overline{x})-(C+P) \int_{0}^{1-\frac{n}{L}} (1-x)f(x)dx = 0 \qquad (5.33)$$

i.e. :

$$\int_{0}^{1-\frac{n}{L}} (1-x)f(x)dx = \frac{1-\bar{x}}{1+\frac{P}{C}} \qquad (5.34)$$

The optimal reservation level L^* is obtained as a solution to expression (5.34) The mean number of passengers turned away with a reservation $\bar{Q}(L)$equals (expression 5.23) :

$$\bar{Q}(L) = \sum_{k=0}^{L-n} (L-n-k)p(k) \qquad (5.35)$$

Since $k = xL$, this is :

$$\bar{Q}(L) = \int_{0}^{1-\frac{n}{L}} \left[L(1-x)-n\right]f(x)dx \qquad (5.36)$$

Based on expressions (5.34) and (5.36) obtained by Gerbracht,[50] the optimal reservation level is determined that realizes maximum revenues, and the mean number of passengers with reservations that cannot find a seat on their desired flight.

5.3. The Optimal Reservation Level as a Function of Time Before Takeoff for a Nonstop Flight and One Air Fare

The optimal reservation level established in previous discussions was not determined as a function of time before takeoff. To this effect, during the entire period starting with the first reservation up until takeoff, the reservation level is constant. The number of reservations rises with time, i.e. is larger and larger as takeoff approaches (Figure 5.1). Since the number of reservations rises with time, models appeared that determine the reservation level as a function of time. It is logical to assume that reservation levels determined as a function of time should decrease over time, due to the rise in the number of reservations over time (Figure 5.2)

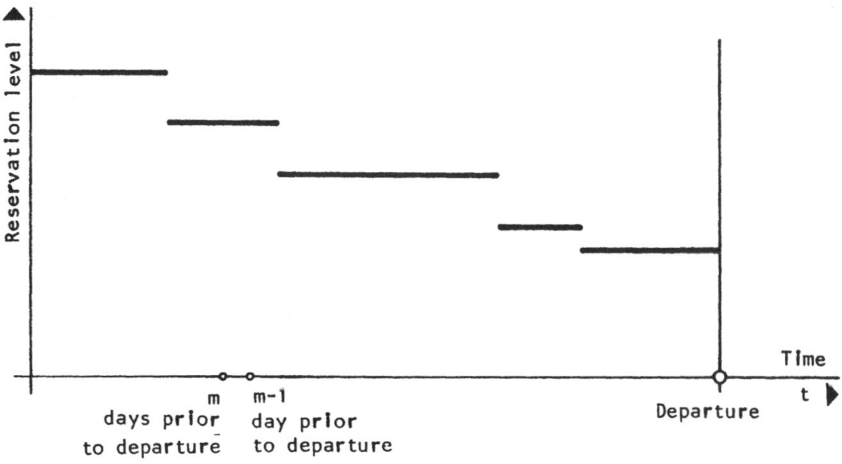

FIGURE 5.2 Decreasing reservation level over time

Shlifer and Vardi [145] developed a model that determines the optimal reservation level as a function of time before takeoff. The basic assumption in this model is that the probability of a reservation being cancelled is independent of the time in which the reservation is made. This assumption was verified by an analysis of statistical data concerning El-Al airline company. The assumption that the probability of a reservation cancellation being independent of the time when the reservation was made was first verified by Martinez and Sanchez [100]. These authors made a very detailed analysis of voluminous statistical material in order to determine the characteristics of passenger behavior during the reservation process. We will briefly present the basic assumptions of the Shlifer and Vardi model.

We denote by $S(t)$ the number of reservations made (the number of seats reserved) at time t before takeoff. With X we denote a random variable representing the number of passengers appearing at takeoff. Based on statistical data collected and processed in the El-Al company, it was shown that this random variable has a normal distribution with mean $S(t)\ a(t)$ and standard deviation $\sqrt{S(t)\cdot b(t)}$, i.e. that :

$$X \sim N(S(t)\cdot a(t),\ \sqrt{S(t)\cdot b(t)})$$

where $a(t)$ and $b(t)$ are parameters that are calculated on the basis of statistical data in the following manner.

First, data must be collected on m flights that have been made on a certain route over a fixed period of time. Let there be $S_1(t)$ reservations t days before departure for the first flight and let there be X_1 passengers that show up for that flight. For the second flight under consideration, let there be $S_2(t)$ reservations t days before takeoff and let there be X_2 passengers that appeared for the flight, etc.

Parameters $a(t)$ and $b(t)$ are obtained as :

$$a(t) = \frac{\sum_{i=1}^{m} X_i}{\sum_{i=1}^{m} S_i(t)} \tag{5.37}$$

$$b(t) = \frac{1}{m} \sum_{i=1}^{m} \frac{\left[X_i - a(t) \cdot S_i(t)\right]^2}{S_i(t)} \tag{5.38}$$

We denote by p^* the greatest allowed probability that the number of passengers on the flight is greater than the number of seats in the aircraft. The number of seats in the aircraft is denoted by n . Thus, the probability that more than n passengers will appear for the flight must be less than or equal to p^*. At time t before takeoff there were $S(t)$ reservations and let the list of reservations be closed from that moment on. The probability that the number of passengers on takeoff is greater than n equals :

$$P(X>n) = P(n<X<\infty) = P\left[\frac{n-S(t)a(t)}{\sqrt{S(t)\ b(t)}} < \right.$$

$$\left. < \frac{X - S(t)a(t)}{\sqrt{S(t)b(t)}} < \infty\right] \tag{5.39}$$

i.e. :

$$P(X>n) = \Phi(\infty) - \Phi\left[\frac{n-S(t)a(t)}{\sqrt{S(t)b(t)}}\right] \tag{5.40}$$

$$P(X>n) = 0,5 + \Phi\left[\frac{S(t)a(t)-n}{\sqrt{S(t)b(t)}}\right] \tag{5.41}$$

whereby :

$$\Phi(t) = \frac{1}{\sqrt{2\pi}} \int_0^t e^{-\frac{u^2}{2}} du$$

is a Laplace function. We also have :

$$\Phi(+\infty) = 0,5; \quad \Phi(-t) = -\Phi(t)$$

Since :

$$P(X > n) = p^* \tag{5.42}$$

then reservation level $S_{p^*}(t)$ for which the probability that the number of passengers at takeoff will be greater than n equals at most the above-given probability p^* , is obtained as the solution to equation :

$$0,5 + \Phi\left[\frac{S(t)a(t)-n}{\sqrt{S(t)b(t)}}\right] = p^* \tag{5.43}$$

We denote by r^* the greatest allowed ratio between the number of passengers turned away with reservations and the number of passengers appearing at takeoff. This ratio is actually the percentage of passengers with regular reservations who cannot get a place on their desired flight. We denote by $T[S(t)]$ the number of passengers turned away for the flight that had $S(t)$ reservations at time t before takeoff and for which the list of reservations was closed as of time t . The number of passengers turned away equals :

$$T[S(t)] = \int_n^{\infty} (x-n) f(x) dx \qquad (5.44)$$

where :

$$f(x) = \frac{1}{\sqrt{S(t)b(t)}\ \sqrt{2\pi}}\ e^{-\frac{[x-S(t)a(t)]^2}{2S(t)b(t)}} \qquad (5.45)$$

is the probability density function of random variable X representing the number of passengers that appeared at takeoff.

By substituting from (5.45) into (5.44) we get :

$$T[S(t)] = \int_n^{\infty} (x-n)\ \frac{1}{\sqrt{S(t)b(t)}\ \sqrt{2\pi}}\ e^{-\frac{[x-S(t)a(t)]^2}{2S(t)b(t)}}\ dx \qquad (5.46)$$

i.e. :

$$T[S(t)] = \int_n^{\infty} \frac{x}{\sqrt{S(t)b(t)}\ \sqrt{2\pi}}\ e^{-\frac{[x-S(t)a(t)]^2}{2S(t)b(t)}}\ dx\ -$$

$$- n \left\{ \Phi(\infty) - \Phi\left[\frac{n-S(t)a(t)}{\sqrt{S(t)b(t)}}\right] \right\} \qquad (5.47)$$

After rearrangement we get :

$$T[S(t)] = \frac{\sigma}{\sqrt{2\pi}} e^{-\frac{[n-S(t)a(t)]^2}{2S(t)b(t)}} +$$

$$+\left[S(t)a(t)-n\right]\left\{0,5+\Phi\left[\frac{S(t)a(t)-n}{\sqrt{S(t)b(t)}}\right]\right\} \qquad (5.48)$$

The ratio between the number of passengers turned away and the expected number of passengers appearing at takeoff equals :

$$r = \frac{T[S(t)]}{S(t)a(t)} \qquad (5.49)$$

By solving the equation :

$$\frac{T[S(t)]}{S(t)\cdot a(t)} = r^* \qquad (5.50)$$

we get the reservation level $S_{r^*}(t)$ for which the largest allowed value of the ratio between the number of passengers turned away and the number of passengers who appeared at takeoff equals r^* .

The reservation level model whose final results are contained in expressions (5.43) and (5.50), i.e,. determining the reservation level based on the maximum allowed probability that the number of passengers at takeoff is greater than the number of seats in the aircraft, and determining the reservation level based on the largest allowed ratio between the number of passengers turned away and the number appearing at takeoff, were developed by Shlifer and Vardi.[145] These authors also proposed to establish the reservation level based on the ratio of the air carrier's losses C_2 when one passenger is turned away and profit C_1 realized by transporting one passenger in order to maximize the air carrier's profits. Shlifer and Vardi proposed to establish the reservation level using values C_1 and C_2 in the following manner.

The mean number of passengers on the plane at departure is denoted by $Y[S(t)]$. The mean number of passengers that appears at takeoff is the sum of the mean number of passengers in the airplane and the mean number of passengers turned away, i.e.

$$S(t)a(t) = Y[S(t)] + T[S(t)] \qquad (5.51)$$

Mean revenues $\bar{R}[S(t)]$ equal :

$$\bar{R}[S(t)] = C_1 \cdot Y[S(t)] - C_2 \cdot T[S(t)] \qquad (5.52)$$

The ratio between losses C_2 when the air carrier turns away one passenger and profit C_1 that the air carrier realizes by transporting one passenger is denoted by q^*, i.e. :

$$q^* = \frac{C_2}{C_1} \qquad (5.53)$$

Reservation level $S(t)$ that maximizes expected revenues $\bar{R}[S(t)]$ is obtained from condition :

$$\frac{d\{\bar{R}[S(t)]\}}{d[S(t)]} = 0 \qquad (5.54)$$

Based on expressions (5.52), (5.53) and (5.54), we get :

$$\frac{\frac{d\{Y[S(t)]\}}{d[S(t)]}}{\frac{d\{T[S(t)]\}}{d[S(t)]}} = q \qquad (5.55)$$

The corresponding value of reservation level $S_{q^*}(t)$ can only be obtained numerically.

Shifler and Vardi [145] also proposed the possibility of determining the reservation level by simultaneously taking into consideration the greatest allowed probability that the number of passengers at departure is greater than the number of seats in the aircraft, the greatest allowed ratio of the number of passengers turned away to the number of passengers appearing at takeoff, while striving to realize greatest possible revenues. This means that the optimal reservation level $S^*(t)$ at time t is obtained as :

$$S^*(t) = min\ \{S_{p^*}(t),\ S_{r^*}(t),\ S_{q^*}(t)\} \qquad (5,55a)$$

whereby $S_{p^*}(t)$, $S_{r^*}(t)$ and $S_{q^*}(t)$ are calculated in the above described manner.

Determining reservation level $L^*(t)$ using expression (5.55a) ensures that the probability that the number of passengers at departure is greater than the number of seats in the aircraft is at most equal to the above-given value p^*. Also, the ratio of the number of passengers turned away to the number appearing at takeoff at most equals the above-given value r^*. The only factor that is not ensured in certain cases when using expression (5.55a) is the air carrier's maximum possible profit.

5.4. The Optimal Reservation Level as a Function of Time Before Departure for a Flight With Two Legs and One Type of Air Fare on Each Leg

The above model to determine the reservation level as a function of time before departure for a nonstop flight with only one air fare was expanded by Shlifer and Vardi [145] to the case of a flight with two legs. Let us note the flight from city A through city B to city C (Figure 5.3).

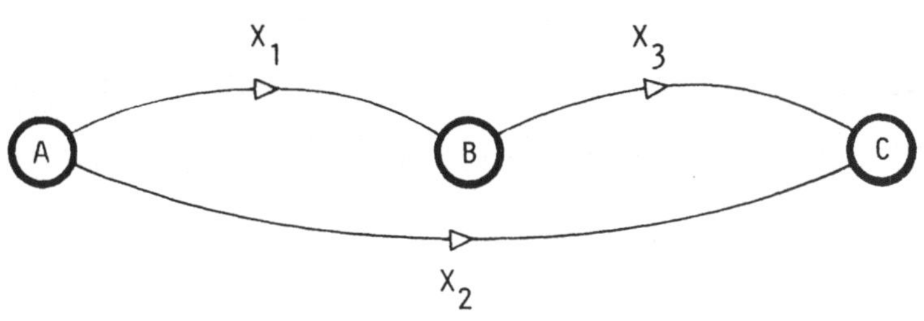

FIGURE 5.3 A flight with two legs

We denote by X_1 the number of passengers that appeared at the airport at city A wanting to travel to city B, and by X_2 the number of passengers appearing at the airport at city A wanting to travel to city C. The number of passengers appearing at the airport at city B wanting to travel to city C is denoted by X_3. We denote by $S_1(t), S_2(t)$ and $S_3(t)$ respectively the number of reservations at time t. As was the case for a nonstop flight, this is :

$$X_1 \sim N(S_1(t)a_1(t), \sqrt{S_1(t)b_1(t)})$$

$$X_2 \sim N(S_2(t)a_2(t), \sqrt{S_2(t)b_2(t)})$$

$$X_3 \sim N(S_3(t)a_3(t), \sqrt{S_3(t)b_3(t)})$$

where $a_i(t)$ and $b_i(t)$, $(i = 1,2,3)$ are corresponding parameters.

When a greater number of passengers than there are seats appears at airport A and/or airport B, i.e. when $X_1+X_2>n$ and/or $X_2 + X_3 > n$ the question arises of setting a reservation level that will achieve the best effects based on certain criteria. Air carriers can strive to minimize the total number of passengers turned away, with each leg transporting the maximum possible number of passengers (in a given situation). Also, one leg might have absolute priority over the others, so that passengers are first turned away who want to travel on a lower priority leg. The first-come-first-served (FCFS) rule might also be applied. In this case, the first passengers registering for the flight get seats.

Let us consider the tactic when we want to minimize the total number of passengers turned away, while maximizing the number of passengers transported on each leg. There are three possible situations :

$$a)\ X_1 + X_2 \leqslant n,\ X_2 + X_3 > n \qquad (5.56)$$

The number of passengers at the airport in city B is greater than the number of seats in the aircraft. The number of passengers to be transported or refused on individual legs so as to minimize the number of refused passengers and maximize the number of passengers transported on each leg are given in Table 5.1.

TABLE 5.1

Leg	*No. of passengers that show up for the flight*	*No. of passengers that get a seat*	*No. of passengers refused*
$A - B$	X_1	X_1	0
$A - C$	X_2	X_2	0
$B - C$	X_3	$n - X_2$	$X_2 + X_3 - n$

$$b)\ X_1 + X_2 > n,\ X_2 + X_3 \leqslant n \qquad (5.57)$$

The number of passengers appearing at airport A is greater than the number of seats in the aircraft. The number of transported and refused passengers on individual legs are given in Table 5.2.

TABLE 5.2.

L e g	No.of passengers that show up for the flight	No.of passengers that get a seat	No. of passengers refused
A - B	X_1	$n - X_2$	$X_1 + X_2 - n$
A - C	X_2	X_2	0
B - C	X_3	X_3	0

$$C)\ X_1 + X_2 > n \quad i \quad X_2 + X_3 > n \qquad (5.58)$$

In this situation the following eight cases are possible :

	L E G		
	A - B	A - C	B - C
1.	$X_1 > n$	$X_2 > n$	$X_3 > n$
2.	$X_1 > n$	$X_2 > n$	$X_3 \leqslant n$
3.	$X_1 > n$	$X_2 \leqslant n$	$X_3 > n$
4.	$X_1 > n$	$X_2 \leqslant n$	$X_3 \leqslant n$
5.	$X_1 \leqslant n$	$X_2 > n$	$X_3 > n$
6.	$X_1 \leqslant n$	$X_2 > n$	$X_3 \leqslant n$
7.	$X_1 \leqslant n$	$X_2 \leqslant n$	$X_3 > n$
8.	$X_1 \leqslant n$	$X_2 \leqslant n$	$X_3 \leqslant n$

In the first and the third cases, passengers on route $A - C$ should not be provided any seats, since this would result in two passengers being refused for each passenger allowed on the plane (one on route $A - B$ and another on route $B-C$). Contrary to this, boarding one passenger on route $A-B$ or $B-C$ only causes one passenger to be refused on route $A-C$. In the second and fourth cases, passengers on route $A-C$ should be provided $n-X_3$ available seats. In the fifth and seventh cases, passengers on route $A-C$ should be provided $n-X_1$ available seats. Finally, in the sixth and eighth cases, passengers on route $A-C$ should be provided $n-X_1$ seats if $n-X_1 > n-X_3$, and $n-X_3$ seats if $n-X_3 > n-X_1$ (since we want to maximize the total number transported on each leg in addition to minimizing the total number of passengers refused). Thus, the number of seats W_2 that should be provided to the X_2 passengers on route $A-C$ equals :

$$W_2 = max\ \{0,\ n-X_1,\ n-X_3\} \quad (5.59)$$

The number of passengers turned away on this leg R_2 will be:

$$R_2 = X_2 - max\ \{0,\ n-X_1,\ n-X_3\} \quad (5.60)$$

Passengers travelling on route $A-B$ are "competing" at airport A for seats with passengers travelling on route $A-C$. Since n available seats must be distributed among $X_1 + X_2$ passengers, we conclude based on expression (5.59) that the number of passengers flying W_1 on route $A-B$ will be :

$$W_1 = n - max\ \{0,\ n-X_1,\ n-X_3\} \quad (5.61)$$

The number of passengers refused R_1 on route $A-B$ is :

$$R_1 = X_1 - n + max\ \{0, n-X_1, n-X_3\} \quad (5.62)$$

Then number of passengers on board and refused on route $B-C$ are, respectively :

$$W_3 = n - max\ \{0,\ n-X_1,\ n-X_3\} \qquad (5.63)$$

$$R_3 = X_3 - n + max\ \{0, n-X_1,\ n-X_3\} \qquad (5.64)$$

The probability that a larger number of passengers will appear at airport A than there are seats in the plane equals :

$$P(X_1+X_2 > n) = 0{,}5+\Phi\left[\frac{S_1(t)a_1(t) + S_2(t)a_2(t)-n}{\sqrt{S_1(t)b_1(t) + S_2(t)b_2(t)}}\right] \qquad (5.65)$$

where $a_1(t)$, $b_1(t)$, $a_2(t)$ and $b_2(t)$ are corresponding parameters.

This probability is calculated in the same manner as the probability defined by expression (5.41). The probability that there will be a greater number of passengers at city B airport than seats in the aircraft equals :

$$P(X_2+X_3 > n) = 0{,}5+\Phi\left[\frac{S_2(t)a_2(t) + S_3(t)a_3(t)-n}{\sqrt{S_2(t)b_2(t)+S_3(t)b_3(t)}}\right] \qquad (5.66)$$

We denote by $p_A{}^*$ the largest allowed probability that the number of passengers at the airport at city A is greater than the number of seats in the aircraft. The corresponding probability for city B airport is denoted by $p_B{}^*$. Reservation levels $S_1^{(p_A^*, p_B^*)}(t)$, $S_2^{(p_A^*, p_B^*)}(t)$ and $S_3^{(p_A^*,\ p_B^*)}(t)$ whose probabilities that a larger number of passengers will appear than seats in the aircraft at airports A and B equal at most the above-given values $p_A{}^*$and $p_B{}^*$, must satisfy the following :

$$0,5 + \Phi\left[\frac{S_1(t)a_1(t) + S_2(t)a_2(t) - n}{\sqrt{S_1(t)b_1(t) + S_2(t)b_2(t)}}\right] = p_A^* \quad (5.67)$$

$$0,5 + \Phi\left[\frac{S_2(t)a_2(t) + S_3(t)a_3(t) - n}{\sqrt{S_2(t)b_2(t) + S_3(t)b_3(t)}}\right] = p_B^* \quad (5.68)$$

The corresponding reservation levels can also be determined based on the greatest allowed values of $r_A{}^*$ and $r_B{}^*$, the ratio of the number of passengers with reservations turned away to the number of passengers appearing at departure at the airports of cities A and B . We denote by $T_A[S_1(t), S_2(t)]$ and $T_B[S_2(t), S_3(t)]$ the number of passengers turned away from airports A and B . Reservation levels $S_1^{(r_A^*, r_b^*)}(t), S_2^{(r_A^*, r_B^*)}(t)$ and $S_3^{(r_A^*, r_B^*)}(t)$ whose ratios of the number of passengers turned away to the number appearing at airports A and B equal the above-given values $r_A{}^*$and $r_B{}^*$, satisfy these equations :

$$\frac{T_A[S_1(t), S_2(t)]}{S_1(t)a_1(t) + S_2(t)a_2(t)} = r_A^* \quad (5.69)$$

$$\frac{T_B[S_1(t), S_2(t)]}{S_2(t)a_2(t) + S_3(t)a_3(t)} = r_B^* \quad (5.70)$$

The estimated number of passengers turned away was presented during the discussion of the reservation policy for a nonstop flight.

5.5. Seat Allocation on a Nonstop Flight With Two Air Fares

Many airline companies have several different types of air fares. With the goal of achieving high load factors, different reductions are given to passengers with fixed departure dates and times, with fixed routes that usually have numerous stopovers, and to passengers with early morning and late evening flights. Reductions are also given to group trips. It has been noted, however (Littlewood[93], Mayer[102], Wang[170], Titze and Griesshaber[164], Rothstein[153]), that passengers with lower fares, no matter what they are based on, usually make their reservations considerably before passengers paying full fare. Business passengers most often pay the full, more expensive fare and reserve their seats considerably later, just several days before the flight. This is because business trips cannot be planned precisely very much in advance. Business passengers are often forced to plan trips just one day in advance and must try to find a seat on their desired flight at the last moment. If the first-come-first-served reservation system is applied, it might happen that a large number of passengers with lower fares make reservations earlier and thereby hinder passengers with higher fares from taking the flight.

A direct result of this is reduced revenues and profit for the air carrier. Although there are usually several different types of fares, for our needs we will assume that there are only two types of fares : the cheap fare and the expensive fare. When there are two air fares, we should clearly not endeavor to maximize the total number of passengers, rather we should maximize revenues. Therefore, the reservation level must be determined for the lower fare so that the air carrier maximizes revenues. In other words, the number of seats in the aircraft must be divided among those with lower and those with higher fares so that revenues are maximized.

We denote by n the number of seats in the aircraft, by L the number of seats reserved for passengers with lower fares and by P the probability that a passenger who would pay the higher fare cannot find a seat because it has been sold to a passenger with a lower fare. A rise in variable L means a rise in probability P (Figure 5.4).

Figure 5.4 presents the probability density function $f(x)$ of the number of passengers paying the higher fare, corresponding to mean μ . A rise in the number of seats for passengers with the lower fare, i.e. a rise in variable L decreases the number of seats for passengers with the higher fare, i.e. variable $n-L$ decreases. A reduction in variable $n-L$ increases the probability that a passenger paying the higher fare cannot find a free seat on his desired flight because it has been sold to a passenger with a lower fare (Figure 5.4). Probability P equals :

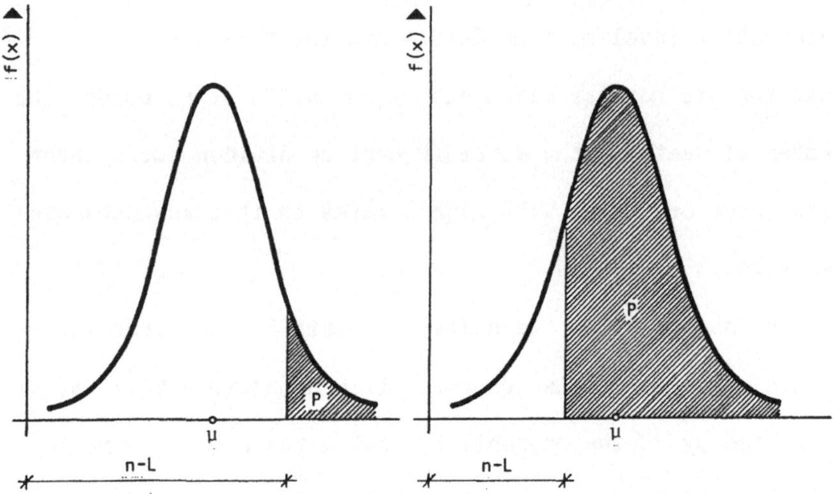

FIGURE 5.4 The influence of variable L on probability P

Littlewood [93] proposed the following way to determine the reservation level L to which reservations are accepted for passengers with the lower fare. We denote by c_1 the revenues from passengers with the lower fare and by c_2 the revenues from passengers with the higher fare. Let L seats be reserved for low fare passengers. Mean revenues per lower fare seat are :

$$1 \cdot c_1$$

Mean revenues per higher fare seat are :

$$(1 - P) \cdot c_2$$

Passengers with lower fares should be accepted until :

$$c_1 \geqslant (1-P)\ c_2 \qquad (5.72)$$

i.e. :

$$(1-P) \leqslant \frac{c_1}{c_2} \qquad (5.73)$$

Reservation level L is determined by solving inequality (5.73). By substituting from (5.71) into (5.73) we get :

$$1 - \int_{n-L}^{\infty} f(x)dx \leqslant \frac{c_1}{c_2} \qquad (5.74)$$

We assume that the random variable representing the number of passengers with higher fares is subject to the normal probability distribution with parameters μ and σ . In this case :

$$1 - \int_{n-L}^{\infty} f(x)dx = 1-\Phi(\infty) + \Phi(\frac{n-L-\mu}{\sigma}) \leqslant \frac{c_1}{c_2} \qquad (5.75)$$

i.e. :

$$\Phi(\frac{n - L - \mu}{\sigma}) \leqslant \frac{c_1}{c_2} - 0{,}5 \qquad (5.76)$$

The main input data to determine reservation level L obtained by solving inequality (5.74), i.e. (5.76) is the probability density function of passengers with higher fares. This function is obtained by analyzing appropriate statistical data from the past. A smaller value for L means that the reservation process is shorter for passengers with the lower fare, and that passengers with higher fares are provided better transportation services. An increase in variable L improves the quality of the transportation services offered to passengers with lower fares.

Broadening the work of Littlewood,[93] Mayer[102] developed a model to determine the reservation level for passengers with lower fares as a function of time before departure. Mayer started with the following assumptions. Passengers paying the lower fare comprise passengers traveling in a group, while those traveling alone were understood to be paying the higher fare. It is also assumed that passengers do not cancel reservations. Thus, it is assumed that each traveler making a reservation will fly and bring the air carrier revenues. The assumption is also introduced that a passenger who is not able to find a free seat on his desired flight will not try to find a place on another flight and is therefore a clear loss to the air carrier. Time period T during which reservations for a certain flight are made is divided into n smaller time periods and denoted respectively by $1,2,3,\ldots,t,$ $\ldots$ n. Mayer[102] also assumed that during each of the time periods $1,2,3,\ldots,t,\ldots,$ n passengers with the lower fare (group trip) reserve their seats before passengers with the higher fare, and that transportation demand in each of these time periods is independent of demand in the periods preceding it. Let time period 1 be the closest and period n the farthest away from takeoff.

Let us introduce the following symbols :

t - time period before departure ($t = 1,2,3,\ldots,n$),

c_1 - revenue from passengers with the lower fare,

c_2 - revenues from passengers with the higher fare,

$p_j(t)$ - the probability that j passengers will appear during time period t paying the lower fare and wanting to make a reservation,

$q_j(t)$ - the probability that j passengers will appear during time period t paying the higher fare and wanting to make a reservation,

n - the number of seats in the aircraft,

$S_c(t)$ - the number of seats during period t for passengers with the lower fare, with c seats available at the beginning of period t ,

$S_c{}^*(t)$ - the optimal number of seats during period t for passengers with the lower fare, with c seats available at the beginning of period t (if this number of seats is achieved revenues are maximized),

$R_c(t)$ - the maximum revenues achieved from the beginning of period t to takeoff, with c seats available at the beginning of period t (revenues are maximized if $S_c{}^*(t)$ seats are offered to the passengers paying the lower fare during every period t).

Let us discuss time period $t=1$ that is closest to departure time. At the beginning of this period, passengers with the lower fare have at their disposition none, one, two,..., or n available seats, i.e. :

$$S_c(t) = 0, 1, 2, \ \ldots \ , n$$

During time period $t=1$ there might appear one, two, three,..., or an infinitely large number of passengers wanting to travel and pay the lower fare. At most $S_c(1)$ passengers paying the lower fare can find seats in the plane. The mean number of passengers paying the lower fare and finding seats in the aircraft equals :

$$\sum_{i=0}^{S_c(1)} i p_i(1) + S_c(1) \sum_{i=S_c(1)+1}^{\infty} p_i(1)$$

The mean revenues from these passengers $R_1[S_c(1)]$ equal :

$$R_1[S_c(1)] = c_1 \sum_{i=0}^{S_c(1)} i p_i(1) + c_1 S_c(1) \sum_{i=S_c(1)+1}^{\infty} p_i(1) \quad (5.77)$$

The air carrier reserves $S_c(1)$ seats in the aircraft during time period $t=1$ for low fare passengers. Let us first consider the case when the low fare passengers were sold $i < S_c(1)$ seats during time $t=1$. The probability of this happening is :

$$\sum_{i=0}^{S_c(1)-1} p_i(1)$$

Mean revenues from high fare passengers $R_2[S_c(1)]$ in this case are :

$$R_2[S_c(1)] = \sum_{i=0}^{S_c(1)-1} p_i(1) \left[\sum_{j=0}^{c-i-1} jq_j(1) + (c-i) \sum_{j=c-i}^{\infty} q_j(1) \right] c_2 \quad (5.78)$$

If low fare passengers are sold all $S_c(1)$ seats, this means that at least $S_c(1)$ of them appeared wanting to travel. The probability of this happening is :

$$\sum_{i=S_c(1)}^{\infty} p_i(1)$$

When low fare passengers are sold $S_c(1)$ seats, mean revenues $R_3[S_c(1)]$ from high fare passengers equal :

$$R_3[S_c(1)] = \sum_{i=S_c(1)}^{\infty} p_i(1) \left[\sum_{j=0}^{c-S_c(1)-1} jq_j(1) + \right.$$

$$\left. + (c-S_c(1)) \sum_{j=c-S_c(1)}^{\infty} q_j(1) \right] c_2 \quad (5.79)$$

Air fare revenues from the beginning of time period $t=1$ to takeoff, with c seats available at the beginning of time

$t=1$ **, directly depend on the number of seats reserved $S_c(1)$ for low fare passengers. Our problem is to determine the value of $S_c(1)$ that maximizes revenues. In other words :**

$$R_c(1) = \max_{S_c(1)=0,1,2,..,c} \left\{ R_1[S_c(1)] + R_2[S_c(1)] + R_3[S_c(1)] \right\} \quad (5.80)$$

i.e. :

$$R_c(1) = \max_{S_c(1)=0,1,2,...,c} \left\{ c_1 \sum_{i=0}^{S_c(1)} ip_i(1) + \right.$$

$$+ c_1 S_c(1) \sum_{i=S_c(1)+1}^{\infty} p_i(1) +$$

$$+ \sum_{i=0}^{S_c(1)-1} p_i(1) \left[\sum_{j=0}^{c-i-1} jq_j(1) + (c-i) \sum_{j=c-i}^{\infty} q_j(1) \right] c_2 +$$

$$+ \sum_{i=S_c(1)}^{\infty} p_i(1) \left[\sum_{j=0}^{c-S_c(1)-1} jq_j(1) + \right.$$

$$\left. + (c-S_c(1)) \sum_{j=c-S_c(1)}^{\infty} q_j(1) \right] c_2 \Bigg\} \quad (5.81)$$

We now determine maximum revenues $R_c(t)$ earned from the beginning of period t to takeoff, with c seats available at the beginning of this period. An analogous expression is :

$$R_c(t) = \max_{S_c(t)=0,1,2,...,c} \left\{ c_1 \sum_{i=0}^{S_c(t)} ip_i(t) + c_1 S_c(t) \cdot \right.$$

$$\cdot \sum_{i=S_c(t)+1}^{\infty} p_i(t) + \sum_{i=0}^{S_c(t)-1} p_i(t) \left[\sum_{j=0}^{c-i-1} g_j(t) \cdot \right.$$

$$\left. \cdot (jc_2 + R_{c-i-j}(t-1)) + (c-i) \sum_{j=c-i}^{\infty} g_j(t) c_2 \right] + \sum_{i=S_c(t)}^{\infty} p_i(t) \cdot$$

$$\cdot \left[\sum_{j=0}^{c-S_c(t)-1} q_j(t) \cdot (jc_2 + R_{c-S_c(t)-j}(t-1)) + (c-S_c(t)) \cdot \right.$$

$$\left. \cdot \sum_{j=c-S_c(t)}^{\infty} q_j(t) c_2 \right] \Bigg\}$$

$$t = 2, 3, \ldots, n-1 \qquad (5.82)$$

The first row of the above sum is the mean revenues during period t from low fare passengers if $S_c(t)$ seats are reserved for them during period t. The addends in the second row of expression (5.82) represent the mean revenues in period t and in following periods from high fare passengers when $S_c(t)$ seats are reserved for low fare passengers during time t and when $i < S_c(t)$ seats are sold to low fare passengers. Finally, the third row of expression (5.82) represents mean revenues during time t and the following periods from high fare passengers when $S_c(t)$ seats are sold to low fare passengers during time period t.

We denote by $S_{c^*}(t)$ the corresponding value from the set of possible values $\{0,1,2,\ldots,c\}$ that $S_c(t)$ can take that maximizes revenues.

After solving the system of equations (5.81) - (5.82), corresponding values are calculated for $R_c(t)$ and $S_{c*}(t)$. The former is calculated as above, with care being taken that at the beginning of period n , all n seats ($c = n$) are available. Variables $S_{c*}(t)$ represents the number of seats in the t periods that must be offered to low fare passengers in order to maximize the air carrier's revenues. The remaining $c - S_{c*}(t)$ seats in any period t should be offered to high fare passengers and as many of them as possible should be sold (in the best case all $c - S_{c*}(t)$ seats can be sold).

The main input data in the above model, which is based on dynamic programming, are the probabilities of a certain number of one or the other type passenger appearing. In order to calculate these probabilities, corresponding statistical data must be previously collected on the past behavior of passengers in order to establish the distribution of the number of travel demands from individual categories of passengers in specific periods of time. Mayer [102] also discussed the case when passengers are allowed to cancel their reservations. In this case, the number of seats for low fare passengers must first be calculated. The remaining number of seats offered to high fare passengers should be treated as the aircraft's capacity and one of the earlier presented models should be used to determine the reservation level.

5.6. Seat Allocation on Flights With Several Legs When There are Several Air Fares

The problem of seat allocation on an aircraft, i.e. the problem of allocating a certain number of seats to passengers with a specific fare, becomes more complex when the flight has several legs. Figure 5.5 schematically presents a flight with two legs.

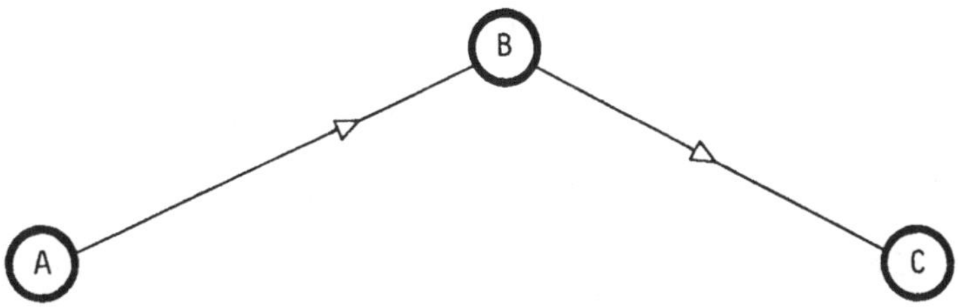

FIGURE 5.5 A flight with two legs

Most often, the average air fare between cities A and C, $\bar{c}_{AC}$ is less than the sum of the fares along legs $A-B$ and $B-C$, i.e. :

$$\bar{c}_{AC} < \bar{c}_{AB} + \bar{c}_{BC} \tag{5.83}$$

We also have :

$$\bar{c}_{AC} > \bar{c}_{AB} \quad \text{and} \quad \bar{c}_{AC} > \bar{c}_{BC} \tag{5.84}$$

We assume that one seat has been sold on route $A-C$. If there are passengers who cannot find a free seat on legs $A-B$ or $B-C$, it is clear that it has been sold to a passenger on route $A-C$, and the air carrier did not earn maximum possible revenues (since $\bar{c}_{AC} < \bar{c}_{AB} + \bar{c}_{BC}$). It is also possible for the air carrier to turn down the passenger on route $A - C$ and sell the seat to a passenger traveling from A to B, expecting for a passenger to appear traveling from B to C. Such passenger might or might not appear. It is also possible to sell the seat on route $B-C$, expecting for a passenger to appear on route $A-B$. The problem of seat allocation becomes particularly complex by the fact that there are several different types of fares on each leg. The following question naturally arises. How many seats should be allocated to passengers in different fare groups between individual pairs of cities so that the air carrier's revenues are maximized?

In order to solve this problem, statistical data must be collected on the number of embarked passengers of certain categories between pairs of cities. Based on these data and bearing in mind the capacity of the aircraft making the flight, the distribution of the number of transportation demands can be obtained. We denote by $X_{i,j}$ a random variable representing the number of transportation demands by passengers wishing to pay fare j when traveling between pairs of cities i.

Wang[170] developed a simple model that allocates seats according to different fares between different pairs of cities. We will explain this model with a simple example. We assume that there are two types of fares (cheap and expensive) for trips between cities $A-B$, $B-C$ and $A-C$ (Figure 5.5). Each seat in the aircraft can be sold in one of the following ways :

1. The passenger travels from A to C and pays the expensive fare (the number of such passengers is a random variable denoted by $X_{1,1}$). We denote the fare by $c_{1,1}$.

2. The passenger travels from A to C and pays the cheap fair ($X_{1,2}$, $c_{1,2}$).

3. The passenger travels from A to B and pays the expensive fare ($X_{2,1}$, $c_{2,1}$). At city B a passenger takes his place and travels to C paying the expensive fare ($X_{3,1}$, $c_{3,1}$).

4. The passenger travels from A to B and pays the expensive fare ($X_{2,1}$, $c_{2,1}$). In city B a passenger takes his place and travels to C and pays the cheap fare ($X_{3,2}$, $c_{3,2}$).

5. The passenger travels from A to B and pays the cheap fare ($X_{2,2}$, $c_{2,2}$). In city B his place is taken by a passenger traveling to C paying the expensive fare ($X_{3,1}$, $c_{3,1}$).

6. The passenger travels from A to B and pays the cheap fare ($X_{2,2}$, $c_{2,2}$). In city B his place is taken by a passenger traveling to C paying the cheap fare ($X_{3,2}$, $c_{3,2}$).

Determination of the probability density function of

random variables $X_{1,1}, X_{1,2}, X_{2,1}, X_{2,2}, X_{3,1}, X_{3,2}$ is obtained from statistical data.

We denote by $S_{i,j}$ the number of seats sold to passengers paying fare i when traveling between cities j. We sell the next seat in the first of the possible six manners, i.e. we sell it to a passenger traveling from A to C paying the expensive fare. The number of sold seats $S_{1,1}$ is increased by one ($S_{1,1} + 1 \rightarrow S_{1,1}$). The remaining numbers of sold seats stay unchanged. Mean revenues R_1 in this case equal :

$$
\begin{aligned}
R_1 = {} & P(X_{1,1} > S_{1,1})c_{1,1} + P(X_{1,2} > S_{1,2})c_{1,2} + \\
& + P(X_{2,1} > S_{2,1})c_{2,1} + P(X_{3,1} > S_{3,1})c_{3,1} + \\
& + P(X_{2,1} > S_{2,1})c_{2,1} + P(X_{3,2} > S_{3,2})c_{3,2} + \\
& + P(X_{2,2} > S_{2,2})c_{2,2} + P(X_{3,1} > S_{3,1})c_{3,1} + \\
& + P(X_{2,2} > S_{2,2})c_{2,2} + P(X_{3,2} > S_{3,2})c_{3,2}
\end{aligned} \qquad (5.85)
$$

Should the seat be sold in another of the possible six fashions, then $S_{1,2}$ increases by $1 (S_{1,2} + 1 \rightarrow S_{1,2})$. The remaining number of sold seats stays unchanged. Value R_2 , as well as values R_3, R_4, R_5 and R_6 are also calculated using expression (5.85). In this manner, mean revenues are calculated for each of the six possible ways of selling a seat. Seats should be sold in the manner that brings the air carrier the greatest mean revenues. These revenues R are :

$$R = max \; \{R_1, R_2, R_3, R_4, R_5, R_6\} \qquad (5.86)$$

As noted by Wang,[170] on flights with greater average load factors, the application of this model can increase realized revenues up to 4%.

Air carrier seat reservation problems have been given considerable attention in the past two decades. Several of the models developed have been successfully implemented. The present situation is such that passengers are overbooked from time to time, but it can be concluded that the number of overbooked passengers compared to the number transported is not very large. The question of financial compensation that the air carrier should offer to overbooked passengers certainly remains open. In its time, Simon's[141] proposal caused considerable polemics in the professional public, referring to both the choice of overbooked passengers that should be taken off the flight, and to the size of the compensation. Simon simply proposed that a type of survey be made of the passengers in order to determine the amount of money that certain passengers were prepared to accept as compensation for not taking their desired flight. Let there be a total of k overbooked passengers. According to Simon, the first k passengers demanding the least compensation should be removed from the flight. Rothstein,[135] who opposed Simon in his own way, indicates after a decade and a half that Simon's proposal was not unfounded, since today, in practice, the employees of many companies negotiate with overbooked passengers by offering them compensations in exchange for being taken off the flight. Nagarajan [107] established that compensations demanded by overbooked passengers are not very high. Vickrey's [167] proposal is also interesting, suggesting that reservations for a flight be purchased without the right of exchange or cancellation. Vickrey proposed for flight reservations to be treated as any other merchandise on the market whose price depends on supply and demand. A passenger wishing to cancel a purchased reservation, according to this proposal, would sell it on the market at market price (the price of the reservation would rise with interest in a specific flight).

Some companies have recently tried to solve the problem of seat allocation or the problem of distributing seats in an aircraft into economy class and first class by introducing partition walls that separate economy class from first class. Depending on the number of reservations in each class, or the number of passengers that appear, this wall is moved to determine the number of seats in each class.

R E F E R E N C E S

1. Abara, J., "Fleet Assignment Using Integer Linear Programming" in Proceedings of the XXIII AGIFORS Symposium, Olive Branch, Mississippi, 1983.

2. Agard, M., "Monthly Assignment of Stewards by Computer", in Proceedings of the X AGIFORS Symposium, Terrigal, Australia, 1970.

3. Anderson, S.E., "Fares and Bookings" in Proceedings of the XII AGIFORS Symposium, Nathanya, Israel, 1972.

4. Antosik, J.L., "Automatic Monthly Crew Assignment : A New Approach", in Proceedings of the XVIII AGIFORS Symposium, Vancouver, Canada, 1978.

5. Arabeyre, J.P., Fearnley, J., Steiger, F.C., Teather, W., "The Airline Crew Scheduling Problem : A Survey", Transportation Science, Vol. 3, No. 2, pp. 140-163.

6. Ashford, N., Stanton, H.P.M., Moore, C.A., Airport Operations, John Wiley and Sons, New York, 1984.

7. Babić, O., Teodorović, D., Tosić, V., "Aircraft Stand Assignment to Minimize Walking", Journal of Transportation Engineering, Vol. 110, No. 1, pp. 55-66, 1984.

8. Babić, O., "Optimizacija opsluzivanja aviona na pristanisnoj platformi aerodroma" (Optimization of Aircraft Ground Handlng Processes at an Airport Terminal) Doctoral Dissertation, Faculty of Transportation and Traffic Engineering, University of Belgrade, 1986.

9. Baker, E.K., Bodin, L.D., Fisher, M., "The Development and Implementation of a Heuristic Set Covering Based System for Air Crew Scheduling", Transportation Policy Decision Making, Vol. 3, pp. 95-110, 1985.

10. Ball, M., Bodin, L., Dial, R., "A Matching Based Heuristic for Scheduling Mass Transit Crews and Vehicles" Transportation Science, Vol. 17, pp. 4-31, 1983.

11. Ball, M., Roberts, A., "A Graph Partitioning Approach to Airline Crew Scheduling", Transportation Science, Vol.19 No. 1 , pp. 107-125, 1985.

12. Bargeton, P., Jacquet, P., "Scheduling of Annual Leaves for Cabin Staff Personnel", in Proceedings of the XV AGIFORS Symposium, Rotorua, New Zealand, 1975.

13. Behbehani, R., Kanafani, A., "Demand and Supply Models of Air Traffic in International Markets", Research Report, ITS - WP - 80 - 5, Institute of Transportation Studies, University of California, Berkeley, 1980.

14. Bennet, B.T., Potts, R.B., "A Rostering Problem in Transportation" in Vehicular Traffic Science - Proceedings of the Third International Symposium on the Theory of Traffic Flow, New York, 1965 (edited by L.C. Edie, R. Herman and R. Rothery), American Elsevier Publishing Company Inc., New York, 1967.

15. Bird, C.J., "A Branch-and-Bound Approach to Aircraft and Maintenance Scheduling", in Proceedings of the VIII AGIFORS Symposium, Bad Kissingen, Germany, 1977.

15a. Bjorkman, B., "Forecast of Intra-European Scheduled Air Passenger Travel 1968-1980", ECAC Doc. No. 3, 1970.

16. Bornemann, D.R., "A Linear Programming Solution to the Reserve Crew Scheduling Problem", in Proceedings of the VIII AGIFORS Symposium, Princeton, New Jersey, 1968.

17. Brown, S.L., Watkins, W.S., "The Demand for Air Travel : A Regression Study of Time Series and Cross-sectional Data in the U.S. Domestic Market", CAB, Washington, D.C. 1968.

18. Buhr, J., "Four Methods for Monthly Crew Assignment - A Comparison of Efficiency", in Proceedings of the VIII AGIFORS Symposium, Vancouver, Canada, 1978.

19. Carter, E.C., Morlok, E.K., "Planning an Air Transport Network in Appalachia", Transportation Engineering Journal of ASCE, Vol. 101, No. TE3, pp. 569-588, 1972.

20. Christofides, N., Korman, S., "A Computational Survey of Methods for the Set Covering Problem", Management Science, Vol. 21, No. 5, pp. 591-599, 1975.

21. Darby-Dowman, K., Mitra, G., "An Extension of Set Partitioning with Application to Scheduling Problems", European Journal of Operational Research, Vol. 21, pp. 200-205, 1985.

22. Deckwitz, T.A., "Interactive Dynamic Aircraft Scheduling" S.M. Thesis, Flight Transportation Laboratory, Department of Aeronautics and Astronautics, Massachusetts Institute of Technology, 1984.

23. De Donnea, F.X., The Determinants of Transport Mode Choice in Dutch Cities, Rotterdam University Press, 1971.

24. De Neufville, R., King, C.R., "Access, Fares, Frequency : Effects on Airport Traffic", Transportation Engineering Journal of ASCE, Vol. 105, pp. 109-125, 1979.

25. Desrosiers, J., Soumis, F., Desrochers, M., Sauve, M., "Routing and Scheduling by Branch-and-Bound on Time Windows", G-83-16, Ecole des Hautes Etudes Commerciales, University de Montreal, 1983 (a).

26. Desrosiers, J., Soumis, F., Desrochers, M., "Routing with Time Windows by Column Generation", G-83-15, Ecole des Hautes Etudes Commerciales, Universite de Montreal, 1983 (b).

27. Desrosier, J., Duman, Y., Soumis, F., "The Multiple Vehicles Many-to-Many Routing Problem with Time Windows", Universite de Montreal, Centre de Recherches sur les Transports - Publication 362, Decembre 1984.

28. Desrosiers, J., Soumis, F., Sauve, M., "Lagrangian Relaxation for Routing with Time Windows", in Operations Research 84 (edited by J.P. Brans), North Holland, 1985.

29. De Vany, A.S., Garges, E.H., "A Forecast of Air Travel and Airport and Airway Use in 1980", Transportation Research, Vol. 6, pp. 1-18, 1972.

30. Douglas, G.W., Miller, J.C., Economic Regulation of Domestic Air Transport : Theory and Policy, the Brookings Institution, Washington, D.C., 1974.

31. English, J.M., Kernan, G.L. "The Prediction of Air Travel and Aircraft Technology to the Year 2000 Using the Delphi Method", Transportation Research, Vol. 10, pp. 1-8, 1976.

32. Eriksen, S., "Policy Oriented Multi-Equation Models of U.S. Domestic Air Passenger Markets", Ph.D thesis, Massachusetts Institute of Technology, 1977.

33. Etcheberry, J., "The Set Covering Problem : A New Implicit Enumeration Algorithm", Operations Research, Vol. 25, No. 5, pp. 760-772, 1977.

34. Etschmaier, M.M., "Projects and Implementation in the Schedule Development Process of an Airline", Technical Report No. 17, Department of Industrial Engineering, University of Pittsburgh, 1973.

35. Etschmaier, M.M., Rothstein, M., "Estimating the Punctuality Rate Inherent in an Airline Schedule", Technical Report no. 19, Department of Industrial Engineering, University of Pittsburgh, 1973.

36. Etschmaier, M.M., Rothstein, M., "Airline Maintenance and the Flight Schedule", Transportation Research Forum, Vol, XV, No. 1, pp. 321-325, 1974.

37. Etschmaier, M.M., Mathaisel, D.F.X., "Aircraft Scheduling the State of the Art" in Proceedings of the XXIV AGIFORS Symposium, Strasbourg, France, 1984.

38. Etschmaier, M.M., Mathaisel, D.F.X., "Airline Scheduling: An Overview", Transportation Science, Vol. 19 , No. 1 , pp. 127-138, 1985.

39. Falkson, L.M., "Airline Overbooking - Some Comments", Journal of Transport Economics and Policy, Vol. 3, No. 3, pp. 252-254, 1969.

40. Fisher, M.L., Jaikumar, R., "A Generalized Assignment Heuristic for Vehicle Routing", Networks, Vol. 11, pp. 109-124, 1981.

41. Gaballa, A., "Planning Aircraft Delays Callout Reserve", in Proceeding of the XI AGIFORS Symposium, Key Biscayne, Florida, 1976.

42. Gagnon, G., "A Model for Flowing Passengers Over Airline Networks", Transportation Science, Vol. 1, No. 3, pp. 232-238, 1967.

43. Gasgo, J.L., "Impersonal Crew Rotation - Rotacyon Mark II", in Proceedings of the XI AGIFORS Symposium, Malaga, Spain, 1971.

44. Gasco, J.L. "Simulation Model for MAD-BCH Air Shuttle", in Proceedings of the XV AGIFORS Symposium, Rotorua, New Zealand, 1975.

45. Gasco, J.L., "Reservations and Booking Control", in Proceedings of the XV AGIFORS Symposium, Bad Kissingen, Germany, 1977.

46. Gasco, J.L., "Report on the Reservations Study Group" in Proceedings of the XX AGIFORS Symposium, New Delhi, India. 1980.

47. Gaudry, M.J.I., Dagenais, M.G., "The Dogit Model", Transportation Research, Vol. 13 B, pp. 105-111, 1979.

48. Gelerman, W., de Neufville, R., "Planning for Satellite Airports", Journal of Transportation Engineering, Vol. 99, NO. TE 3, pp. 537-551, 1972.

49. Gerbracht, R., "A New Algorithm for Very Large Crew Pairing Problems", in Proceedings of the XVIII AGIFORS Symposium, Vancouver, Canada, 1978.

50. Gerbracht, R., "Optimum Overbooking Levels", in Proceedings of the XIX AGIFORS Symposium, Pugnochiuso, Italy, 1979.

51. Gerbracht, R., "A Model for Multi-Fleet Aircraft Routing", Paper presented at the ORSA/TIMS Joint National Meeting, 1982.

52. Gerbracht, R.,"Third Generation Pairing Planning System", in Proceedings of the XXIII AGIFORS Symposium, Olive Branch, Mississippi, 1983.

53. Gertsbach, J. Gurevich, Yu., "Constructing an Optimal Fleet for a Transportation Schedule", Transportation Science, Vol. 11, No. 1, pp. 20-36, 1977.

54. Ghobrial, A.A., "Analysis of the Air Network Structure : The Hubbing Phenomenon", Ph.D thesis, University of California, Berkeley, 1983.

55. Giannessi, F., Nicoletti, B., "The Crew Scheduling Problem : An Interactive System Approach", in Proceedings of the XVII AGIFORS Symposium, Bad Kissingen. Germany, 1977.

56. Gilett, B., MIller, L., "A Heuristic Algorithm for the Vehicle Dispatch Problem", Operations Research, Vol. 22, pp. 340-346 , 1974.

57. Ganert, W., "A 'Timetable' Approach to the Assignment of Pilots to Rotations", in Proceedings of the XXIV AGIFORS Symposium, Strasbourg, France, 1984.

58. Gordon, S., de Neufville, R., "Design of Air Transportation Networks", Transportation Research, Vol. 7, pp. 207-222, 1973.

59. Griesshaber, R., "A Heuristic Model for the Impersonal Phase of Crew Scheduling", in Proceedings of the VIII AGIFORS Symposium, Princeton, New Jersey, 1968.

60. Griesshaber, R., "Two Models for Crew Scheduling", in Proceedings of the XI AGIFORS Symposium, Malaga, Spain, 1971.

61. Griesshaber, R., "Model for Crew Scheduling for a Short-Haul Fleet", in Proceedings of the XX AGIFORS Symposium, New Delhi, India, 1980.

62. Grlihes, D., "Aerodrom Bor" (Bor Airport). Graduation Thesis, Faculty of Transportation and Traffic Engineering, Belgrade, 1977.

63. Grulović, M., "Organizacije najekonomicnijeg nacina vazdusnog transporta na linijama Beograd-London i Zagreb-London" (Organizing the Most Efficient Air Transport on the Belgrade-London and Zagreb-London Routes), Graduation Thesis, Faculty of Transportation and Traffic Engineering, 1979.

64. Haydon, B., "Aircrew Standby Policies for Long Haul Carriers", in Proceedings of the XIII AGIFORS Symposium,

65. Holst, O., Sorensen, B., "Combined Scheduling and Maintenance Planning for an Aircraft Fleet", in Operational Research '84 (edited by J.P. Brans), North-Holland, 1985.

66. Hwang, C.L., Yoon, K., Multiple Attribute Decision Making Springer-Verlag, Berlin, 1981.

67. Hyman, W., Gordon. L., "Commercial Airline Scheduling Technique", Transportation Research, Vol. 2, pp. 23-29.

68. Jaw, J.J., Odoni, A.R., Psaraftis, H.N., Wilson, N.H.M., "A Heuristic Algorithm for the Multi-Vehicle Advance Request Dial-A-Ride Problem with Time Windows", Working Paper MIT-UMTA-83-L, Massachusetts Institute of Technology, 1984.

69. Johnston, E.E., Jones, V.J., Ritchie, B.J.R., "Measuring Consumer Perceptions of Airline Competition", paper presented at the World Conference on Transport Research, London, 1980.

70. Kanafani, A., Sadoulet, E., "The Partitioning of Long Haul Air Traffic - A Study in Multinomial Choice", Transportation Research, Vol. 11, pp. 1-8, 1977.

71. Kanafani, A., Chang, P., "A System Analysis of Short Haul Air Transportation", Research Report, Institute of Transportation Studies, University of California, Berkeley, 1979.

72. Kanafani, A., "Aircraft Technology and Network Structure in Short-Haul Air Transportation", Transportation Research, Vol. 15A, pp. 305-314.

73. Kanafani, A., Ghobrial, A., "Aircraft Evaluation in Air Network Planning", Transportation Engineering Journal of ASCE, Vol. 108, No. TE 3, pp. 282-300, 1982.

74. Kanafani, A., Transportation Demand Analysis, McGraw-Hill Book Company, New York, 1983.

75. Kanafani, A., Ghobrial, A.A., "Airline Hubbing - Some Implications for Airport Economics", Transportation Research, Vol. 19A, No. 1, pp. 15-27, 1985.

76. Kayukawa, K., "Fleet Simulation and a Queuing Model Applied to Aircraft Maintenance Planning", in Proceedings of the XVII AGIFORS Symposium, Bad Kissingen, Germany, 1977.

77. Kedic, M., "Pregled postojecih teorijskih models raspodele putnika po vidovima prevoza i analiza faktora koji opredeljuju putnike za koriscenje aviona na poslovnim relacijama" (Review of Theoretical Passenger Modal Split Models and Analysis of Factors that Induce Passengers to Fly on Business Trips), Graduation Thesis, Faculty of Transportation and Traffic Engineering, Belgrade, 1984.

78. Khasnabis, S., Cynecki, M.J., Flak, M.A., "Systemic Calibration of Multinomial Logit Models", Journal of Transportation Engineering, Vol. 109, No. 2, pp. 209-231, 1983.

79. Kirby, M.G., "Optimal Pricing Policies for Air Transport Networks : Some Comments on Economic Regulation - and the Australian Domestic Air Industry", Transportation Research, Vo. 14A, No. 4, pp.275-278, 1980.

80. Konig, P., "Short Term Aircraft Assignment" in Proceedings of the XVI AGIFORS Symposium, Key Biscayne, Florida, 1976.

81. Koppelman, F.S., Pas, E.I., "Models of Consumer Travel Choice Behavior", Paper presented at the 59th Annual Meeting of the Transportation Research Board, Washington, D.C., 1980.

82. Kraft, J., Kraft, A., "Empirical Estimation of the Value of Travel Time Using Multi Mode Choice Models", Journal of Econometrics, Vol. 1, pp. 317-326, 1974.

83. Kuhn, H.W., "The Hungarian Method for the Assignment Problem", Naval Research Logistic Quarterly, Vol. 2, pp. 83-97, 1955.

84. Labombarda, P., Nicoletti, B., "Aircraft Rotations by Computer", in Proceedings of the XI AGIFORS Symposium, Malaga, Spain, 1971.

85. Ladany, S.P., Hersh, M., "Non-Stop VS One Stop Flights", Transportation Research, Vol. 11, No.3, pp. 155-159,1977.

86. Lauchli, U., "Dynamic Fleet Planning : A Quick and Approximate Evaluation of Many Scenarios", in Proceedings of the XIX AGIFORS Symposium, Pugnochiuso, Italy, 1979.

87. Lave, C.A., "A Behavioral Approach to Modal Split Forecasting", Transportation Research, Vol. 3, pp. 463-480, 1969.

88. Lee, K., Reno, R., "Aircraft Maintenance Scheduling", in Proceedings of the XVIII AGIFORS Symposium, Vancouver, Canada, 1978.

89. Lemke, C.E., Salkin, H.M., Spielberg, K., "Set Covering by Single-Branch Enumeration with Linear-Programming Subproblems", Operations Research, Vol. 19, No. 4, pp. 998-1022, 1971.

90. Levin, A., "Scheduling and Fleet Routing Models for Transportation Systems", Transportation Science, Vol. 5, No. 3, pp. 232-255, 1971.

91. Levin, I.P., Herring, R.D., "Functional Measurement of Qualitative Variables in Mode Choice: Ratings of Economy, Safety and Desirability of Flying Versus Driving", Transportation Research, Vol. 15A, pp. 207-214, 1981.

92. Littlewood, K., "Aircrew Flight Sequencing Methods for Planning and Bid Line Construction",in Proceedings of the X AGIFORS Symposium, Terrigal, Australia, 1970.

93. Littlewood, K., "Forecasting and Control of Passenger Bookings", in Proceedings of the XII AGIFORS Symposium, Nathanya, Israel, 1972.

94. Lof, A.M., "A Branch-and-Bound Algorithm for Determining the Minimal Fleet Size of a Transportation System", Transportation Science, Vo. 4, NO. 2, pp. 159-163, 1970.

95. Maimon, A., Peterson, R.M., "Applications of an Airline Fleet Assignment Model", Paper presented at the Joint Meeting ORSA/TIMS, Las Vegas, Nevada, 1975.

96. Marchettini, F., "Automatic Monthly Cabin Crew Rostering Procedure", in Proceedings of the XX AGIFORS Symposium, New Delhi, India, 1980.

97. Marsten, R.E., "An Algorithm for Large Set Partitioning Problems", Management Science. Vol. 20, NO. 5, pp. 774-787, 1974.

98. Marsten, R.E., Muller, M.R., Killion, C.I., "Crew Planning at Flying Tiger : A Successful Application of Integer Programming", Management Science, Vol.s 25, No. 12, pp. 1175-1183, 1979.

99. Marsten, R.E., Shepardson, F., "Exact Solution of Crew Scheduling Problems Using the Set Partitioning Model : Recent Successful Applications", Networks, Vol. 11, pp. 165-177, 1981.

100. Martinez, R., Sanchez, M., "Automatic Booking Level Control", in Proceedings of the X AGIFORS Symposium, Terrigal, Australia, 1970.

101. Mathaisel, D.F.X., "Fleet Assignment with Variable Demand : A Goal Programming Approach", in Proceedings of the XXIII AGIFORS Symposium, Olive Branch, Mississippi, 1983.

102. Mayer, M., "Seat Allocation, or a Simple Model of Seat Model Allocation via Sophisticated Ones", in Proceedings of the XVI AGIFORS Symposium, Key Biscayne, Florida, 1976.

103. McFadden, D., "The Theory and Practice of Disaggregate Demand Forecasting for Various Modes of Urban Transportation", Working Paper No. 7623, Institute of Transportation Studies, University of California, Berkeley, 1976.

104. Miller, J.C., "A Time-of-Day Model for Aircraft Scheduling", Transportation Science, Vol. 6, No. 3, pp. 221-246, 1972.

105. Mitchell, J.S., "Goal Programming for Scheduling Reserve Pilot Availability", in Proceedings of the XVII AGIFORS Symposium, Bad Kissingen, Germany, 1977.

106. Mondria, H.R., "An Algorithm for the Long-Haul Crew Rotation Problem", in Proceedings of the XI AGIFORS Symposium, Malaga, Spain, 1971.

107. Nagarajan, K.V., "On an Auction Solution to the Problem of Airline Overbooking", Transportation Research, Vol. 13A, No. 2, pp.111-114, 1979.

108. Nicolaidis, G.C., "Quantification of the Comfort Variable",Transportation Research, Vol. 9, pp. 55-66, 1975.

109. Nicoletti, B., "Automatic Crew Rostering", Transportation Science, Vol. 9, No. 1, pp. 33-42, 1975.

110. Niederer, M., "Frequency and Schedule Optimization by LP", in Proceedings of the XI AGIFORS Symposium, Malaga, Spain, 1971.

111. Odier, E., Lascaux, F., Hie, H., "Medium Haul Trip Pairing Optimization", in Proceedings of the XXIII AGIFORS Symposium, Olive Branch, Mississippi, 1983.

112. Pierce, J.F., "Application of Combinatorial Programming to a Class of All-Zero-One Integer Programming Problems" Management Science, Vol. 15, No. 3, pp. 191-209, 1968.

113. Pierce, J.F.,, "Special Survey Paper - Direct Search Algorithms for Truck-Dispatching Problems", Transportation Research, Vol. 3, pp. 1-42, 1969.

114. Pierce, J.F., Lasky, J.S., "Improved Combinatorial Programming Algorithms for a Class of All-Zero-One Integer Programming Problems", Management Science, Vol. 19, No. 5, pp. 528-543, 1973.

115. Pollack, M., "An Aircraft Scheduling Model for Airline Planning", presented at the 41st National Meeting of ORSA, New Orleans, Louisiana, April 26-28, 1972.

116. Pollack, M., "Some Elements of the Airline Fleet Planning Problem", Transportation Research, Vol. 11, pp. 301-310, 1977.

117. Pollack, M., "Measures of Airline Activity and Efficiency : A Descriptive Survey", Transportation Research, Vol. 13A, No. 5, pp. 345-349, 1979.

118. Pollack, M., "Airline Route-Frequency Planning : Some Design Trade-Offs", Transportation Research, Vol. 16A, No. 2, pp. 149-159, 1982.

119. Potts, R.B., "Design and Analysis of an In-Flight Passenger Survey", in Proceedings of the X AGIFORS Symposium, Terrigal, Australia, 1970.

120. Potts, R.B., Oliver, R.M., Flows in Transportation Networks, Academic Press, 1972.

121. Powell, W.B., "Analysis of Airline Operating Strategies Under Stochastic Demand", Transportation Research, Vol. 16B, No. 1, pp. 31-43, 1982.

122. Psaraftis, H.N., "An Exact Algorithm for the Single Vehicle Many-to-Many Dial-A-Ride Problem with Time Windows", Transportation Science, Vol. 17, No. 3, pp. 351-359, 1983.

123. Quandt, R.E., Baumol, W.J., "The Demand for Abstract Modes: Theory and Measurement", Journal of Regional Science, Vol. 6, No. 2, pp. 13-26. 1966.

124. Quandt, R.E., "The Theory of Travel Demand", Transportation Research, Vol. 10,. pp. 411-413, 1976.

125. Rallis, T., Intercity Transport, The MacMillan Press, London, 1977.

126. Renard, G., "Competition in Air Transportation : An Econometric Approach", Thesis presented to the Massachusetts Institute of Technology, Cambridge, Massachusetts in 1970, in partial fulfillment of the requirements for the degree of Master of Science.

127. Richardson, R., "An Optimization Approach to Routing Aircraft", Transportation Science, Vol. 10, No. 1, pp. 52-71, 1976.

128. Richter, R.J., "Optimal Aircraft Rotations Based on Optimal Flight Timing", in Proceedings of the VIII AGIFORS Symposium, 1968.

129. Richter, R.J. "Experience with the Aircraft Rotation Model", in Proceedings of the X AGIFORS Symposium, Terrigal, Australia, 1970.

130. Roberts, F.S. Applied Combinatories, Prentice-Hall Inc., Englewood Cliffs, New Jersey, 1984.

131. Rothstein, M., Stone, W., "Passenger Booking Levels", in Proceedings of the VII AGIFORS Symposium, 1967.

132. Rothsteien, M., "An Airline Overbooking Model", Transportation Science, Vol. 5, pp. 180-192, 1971a.

133. Rothstein, M., "Airline Overbooking: The State of the Art", Journal of Transport Economics and Policy, Vol. 5, No. 1, pp. 96-99, 1971b.

134. Rothstein, M. "Airline Overbooking " Fresh Approaches Are Needed", Transportation Science, Vol. 9, pp.169-173, 1975.

135. Rothstein, M., "OR and the Airline Overbooking Problem", Operations Research, Vol. 33, No. 2, pp. 237-248, 1985.

136. Rubin, J., "Technique for the Solution of Massive Set Covering Problems, with Application to Airline Crew Scheduling", Transportation Science, Vol. 7, No. 1, pp. 34-48, 1973.

137. Sasaki, T., "Probabilistic Models for Trip Distribution" in Proceedings of the Fourth International Symposium on the Theory of Traffic Flow, Karlsruhe, 1968.

138. Scarbrough, J.E., "A Mixed-Integer Programming Model for the Scheduling of Aircraft by Time of Day Within a Network of Airports", Paper presented at the TIMS/ORSA Joint National Meeting, 1977.

139. Scheid, F., Numerical Analysis, McGraw Hill Book Company, New York, 1968.

140. Sen, A., Soot, S., Pagitsas, E., "The Logit Modal Split Model : Some Theoretical Considerations, Transportation Research, Vol. 12, pp. 321-324, 1978.

141. Simon, J.L. "An Almost Practical Solution to Airline Overbooking", Journal of Transport Economics and Policy, Vol. 2, No. 2, pp. 201-202, 1968.

142. Simon, J.L., "Airline Overbooking - A Rejoinder", Journal of Transport Economics and Policy, Vol. 4, No. 2, pp. 212-213, 1970.

143. Simon, J.L., "Airline Overbooking " The State of the Art - A Reply", Journal of Transport Economics and Policy, Vol. 6, NO. 3, pp. 254-256, 1972.

144. Sherali, H.D., Rios, M., "An Air Force Crew Allocation and Scheduling Problem", Journal of Operational Research Society, Vol. 35, No. 2, pp. 91-103, 1984.

145. Shlifer, E., Vardi, Y., "An Airline Overbooking Policy", Transportation Science, Vol. 9, pp. 101-114, 1975.

146. Simpson, R.W., "A Review of Scheduling and Routing Models for Airline Scheduling", in Proceedings of the IX AGIFORS Symposium, Broadway, England, October, 1969.

147. Smith, C.M., Kyle, J.S., "On-Line Flight Schedule Development",in Proceedings of the XII AGIFORS Symposium Nathanya, Israel, 1972.

148. Sommers, A.N., "A Technique for Estimating V/STOL Market Shares", Transportation Science, Vol. 3, No. 1, pp. 1-7, 1969.

149. Soumis, F., Ferland, J.A., Rousseau, J.M., "MAPUM : A Model for Assigning Passengers to a Flight Schedule", Transportation Research, Vol. 15A, pp. 155-162, 1981.

150. Stojkovic, G., "Pregled postojecih teorijskih models i analiza faktora koji opredeljuju putnike za koriscenje aviona na turistickim relacijama" (Review of Theoretical Models and Analysis of Factors Inducing Passengers to Fly on Tourist Routes), Graduation Thesis, Faculty of Transportation and Traffic Engineering, Belgrade, 1984.

151. Swan, W.M., "A Systems Analysis of Scheduled Air Transportation Networks",Report FTL-R79-5, Massachusetts Institute of Technology, 1979.

152. Taha, H., Operations Research, MacMillan Publishing Company, Inc., New York, 1982.

153. Taneja, N.K., The Commercial Airline Industry, Lexington Books, Lexington, Massachusetts, 1978.

154. Teodorović, D., Acimovic-Raspopovic, V., "Primena entropijskog modela srednje duzine putovanja na ocenu intenzitet putnickih tokova u jugoslovenskom unutrasnjem vazdusnom saobracaju" (Application of the Entropic Model of Medium Haul Flights to Estimate Passenger Flows in Yugoslav Domestic Air Traffic), Collection of Papers from the Faculty of Transportation and Traffic Engineering, no. 5, pp. 123-133, 1981.

155. Teodorović, D., "Flight Frequency Determination", Journal of Transportation Engineering, Vol. 109, No. 5, pp. 747-757, 1983.

156. Teodorović, D., Tosić, V., Jovanović, T., "Prognoza putnickih tokova u domacem vazdusnom saobracaju na linijama is SR Slovenije" (Passenger Flow Forecast for Domestic Flights from SR Slovenia), Saobracaj, year XXXI, no. 2, pp. 10-21, 1984a.

157. Teodorović, D., Guberinić, S., "Optimal Dispatching Strategy on an Airline Network After a Schedule Perturbation", European Journal of Operational Research, Vol. 15, No. 2, pp. 178-182, 1984b.

158. Teodorović, D., Babić, O., "Primena Levinovog algoritma na vrsenje rotacije aviona flote "JAT"-a" (Applying Levin's Algorithm to Rotate Aircraft from JAT's Fleet), Proceedings from SYM OP IS '85, Herceg Novi, 1985a.

159. Teododrović, D., "Odredjivanje frekvencija letenja na mrezi linija u slucaju postojanja konkurencije" (Determining Flight Frequency on a Route With Competition), Proceedings from SYM OP IS '85, Herceg Novi, 1985b.

160. Teodorović, D., "A Model for Designing the Meteorologically Most Reliable Airline Schedule", European Journal of Operational Research, Vol. 21, No. 2, pp. 156-164, 1985c.

161. Teodorović, D., "Multicriteria Ranking of Air Shuttle Alternatives", Transportation Research, Vol. 19B, No. 1, pp. 63-72, 1985d.

162. Teodorović, D., "Multiattribute Aircraft Choice for Airline Networks", Journal of Transportation Engineering Vol. 112, No. 6, pp. 634-646, 1986a.

163. Teodorović, D., Transportation Networks, Gordon and Breach, London, 1986b.

164. Titze, B., Griesshaber, R., "Realistic Passenger Booking Behavior and the Simple Low-Fare/High-Fare Seat Allotment Model", in Proceedings of the XXIII AGIFORS Symposium, OLive Branch, Mississippi, 1983.

165. Tomlin, J.A., Tomlin, S.G., "Traffic Distribution and Entropy", Nature, Vol. 220, pp. 974-976, 1968.

166. Tosić, V., Babić, O. "Primer detaljne analize upotrebljivosti aerdroma u odnosu na meteo uslove" (Detailed Analysis of Airport Usability as a Function of Meteorological Conditions), Proceedings of the XI Congress of the Yugoslav Road Association, Opatija,1982.

167. Vickrey, W. "Airline Overbooking : Some Further Solutions", Journal of Transport Economy and Policy, Vol. 6, No. 3, pp. 257-270, 1972.

168. Vukadinović, S., "Elementi teorije verovatnoce i matematicke statistike" (Elements of the Theory of Probability and Mathematical Statistics), Privredni Pregled, Belgrade, 1978.

169. Vukadinović, S., Popović, J., Teodorović, D. Zbirka resenih zadataka iz matematicke statitike (Collection of Solved Mathematical Statistics Problems), Faculty of Transportation and Traffic Engineering, Belgrade, 1981.

170. Wang, K., "Optimum Seat Allocation for Multi-Leg Flights with Multiple Fare Types", in Proceedings of the XXIII AGIFORS Symposium, Olive Branch, Mississippi, 1983.

170a. Williamson, W.G., "Computer Programs for Fleet and Schedule Planning", in Proceedings of the VII AGIFORS Symposium, 1967.

171. Young, D.R., "Scheduling a Fixed-Schedule, Common Carrier Passenger Transportation System", Transportation Science, Vol. 4, No. 3, pp. 243-269, 1970.

I N D E X